HOUSTON, WE HAVE
A NARRATIVE

HOUSTON, WE HAVE A NARRATIVE

Why Science Needs Story

RANDY OLSON

The University of Chicago Press

Chicago and London

Randy Olson was a tenured professor of marine biology at the University of New Hampshire before moving to Hollywood and entering film school at the University of Southern California. He has written and directed a number of films, including the acclaimed *Flock of Dodos*, and he is the author of numerous successful books, including *Don't Be Such a Scientist*.

The University of Chicago Press, Chicago 60637
The University of Chicago Press, Ltd., London
© 2015 by Randy Olson
All rights reserved. Published 2015.
Printed in the United States of America

24 23 22 21 20 19 18   5

ISBN-13: 978-0-226-27070-8 (cloth)
ISBN-13: 978-0-226-27084-5 (paper)
ISBN-13: 978-0-226-27098-2 (e-book)

DOI: 10.7208/chicago/9780226270982.001.0001

Library of Congress Cataloging-in-Publication Data

Olson, Randy, 1955– author.
Houston, we have a narrative : why science needs story / Randy Olson.
pages ; cm
Includes bibliographical references and index.
ISBN 978-0-226-27070-8 (cloth : alk. paper)—ISBN 978-0-226-27084-5 (pbk. : alk. paper)—ISBN 978-0-226-27098-2 (ebook)   1. Communication in science. 2. Science in motion pictures. 3. Storytelling.   I. Title
Q223.O47 2015
501'.4—dc23
2015017768

♾ This paper meets the requirements of
ANSI/NISO Z39.48-1992 (Permanence of Paper).

Contents

Why Science Needs Story

Science is permeated with story. Both the scientific method and the communication of science are narrative processes. Yet the power and structure of story are neither widely taught nor openly advocated. Science is now facing significant problems stemming from this oversight, from the proliferation of false positives within the field to a growing antiscience sentiment outside the field. Help is needed, but the experts in the humanities who ought to provide assistance are buried in their own problems and lack a practical perspective. I argue that science should turn to the people who have spent a century learning and applying the real world power of narrative—the writers, directors, actors, editors and other veterans of Hollywood. There is nothing to be feared from narrative. It pervades all aspects of human culture. Scientists must realize that science is a narrative process, that narrative is story, therefore science needs story.

I

INTRODUCTION

WHY SCIENCE NEEDS STORY

Houston, We Have a Narrative

"How would you like to share your communication ideas with an audience of 1,000 eager minds?"

That was my friend Megan's invitation, asking me to take part in a panel discussion at a 2013 meeting of ocean scientists in San Diego. It's the sort of activity I do these days. I used to be a scientist, I became a filmmaker, and now I work with scientists helping them communicate more effectively with the public. I could hear the excitement in her voice—it was a chance to present my work on communication and storytelling to a large and interested crowd. It sounded good, so I agreed.

As the summer went by I didn't give it much thought, then about six weeks before the event I looked at the meeting website to see what I had signed up for. There were two other panelists, both of whom I know and who are more than ten years my senior. But more importantly, they are two of the world's top experts on the subject of sea level rise—something I know virtually nothing about. Furthermore, looking at the title of the panel, "Responding to Sea Level Rise," there was no clue where I, the scientist-turned-filmmaker, fit in. The event felt like "Two Great Scientists (plus this other guy)."

I said to myself, "Houston, we have a problem."

I called Megan and asked if there was a reason why she had put me on a panel for something I know so little about. She said, "Yes, yes, yes—these guys are dying to work with you. They want you to use your storytelling knowledge to do makeovers on their presentations."

We talked it through. By the end I understood her idea and it sounded cool—a chance to implement the teachings of my books and workshops on the need to tell better stories. Great!

I set to work writing an email to the four of us, laying out my initial ideas. I would reshape the scientists' material into a set of stories they and I would tell, taking turns presenting different parts. It seemed perfect . . . until the scientists replied.

There was immediate pushback. One of them said that his presentation was already set—he had been giving it for over a year—everyone loves it. Basically, it's not broken, no need to fix it, thanks. The other was in Europe and said he didn't have the time for changes.

I pushed a little harder, explaining my ideas further, including how the team presentation style would add energy to the normally dull panel format. They didn't seem to like my labeling things as "normally dull." And did I mention they were ages 68 and 70?

"We just don't need it," one of them wrote. But of course I ignored that. I was still sold on Megan's enthusiasm, so I did what I always do—I kept pushing. Finally the truth started coming out.

"Look," one of them replied, "both of us are known as good speakers. We're very busy. We'll show up and give our standard talks. It will be fine."

I shot back, "I know, but what I want is more than 'fine.' With the power of narrative we can reach a higher level and give the crowd an event to remember."

"I just don't see how it's going to work," his next email said. "You're talking about us taking repeated turns speaking. We'll be getting up and sitting down, bumping into each other—it sounds like a mess."

I replied, "No, trust me, the audience will appreciate the energy of the team effort. It shows we're listening to each other."

And then . . . well, there were a couple more exchanges, until one of them finally said, "Randy, all of us have given countless numbers of these talks. We *all* know how they work. We *all* have the same amount of experience. There's just no need for what you're describing."

And that was it. A moment of realization for me.

Presentations given by scientists, administrators, students—pretty much anyone—are very, very personal. They are an extension of the speaker's inner being, an expression of the ego. In this age of TED Talks, everyone is working on their presentations—running them by their friends and family, honing and shaping them. My asking to get in and mess with others' presentations is like asking to come over and reorganize their underwear drawers. It really is that personal.

I could sense I had hit the limit. An eruption was approaching. Which meant it was time to end it by showing how hopeless the predicament was. I did this by tossing a hand grenade into the discussion so there would be no lingering doubts.

Drawing on my most condescending tone, I replied, "Eh hem . . . only one of us has over two decades of mass communications experience . . ."

I hit SEND and waited less than two minutes for the nuclear missile I knew would come back, which it did, in the form of a short email that began,

"Well, Randy . . . . aren't we special. I suggest you check yourself before this entire event unravels."

There was more to it that was even worse. I sat there looking at my computer screen thinking, "Whoa . . . ," and figured that was enough. I didn't reply. Instead I was breathing deeply as I headed out the door for a cooling-off jog.

I thought about what I was trying to do. These two guys were the sources of knowledge—they were the ones who actually knew something truthful about the real world. I was this horrible agent of conformity wanting to reshape their words and information, to transform the real world into the narrative world.

This same shaping process happened with the iconic quote from the Apollo 13 mission to the moon. The original words spoken by astronaut Jack Swigert in 1970 when an oxygen tank exploded on board were "Houston, we've *had* a problem *here*." But 25 years later, when Tom Hanks delivered the line in the movie version of the events, the words were "Houston, we *have* a problem."

What changed and why? Two things. The Hollywood folks made the line more concise (fewer words) and they made it more compelling (present tense makes it more urgent). I wanted to do this with the scientists—keep things accurate yet make them conform better to the constraints of the narrative world in which we live.

But this sort of text manipulation worries scientists. They want people to know how things are in the real world, and they dream of simply being able to "see it, say it." They want to tell you the truth, exactly as they see it, without having to rearrange anything, because the rearranging process can be dangerous. Rearranging things comes with risks—at the mildest just getting it wrong, at worst deceiving people.

But the problem is, "see it, say it" doesn't work. Not even in the world of science, as Nobel laureate P. B. Medawar first addressed in the 1960s with his essay "Is the Scientific Paper a Fraud?" He agonized over the transition that must take place, where scientists have to give in to a third step, ending up with "see it, shape it, say it." This is what scientists do every day in the process of editing their scientific papers.

Yet the strange thing is that, despite having made major concessions over the past century to this need to shape things, scientists still have little awareness of it. Let me tell you about a little experiment I've run to demonstrate this lack of awareness.

IMRADical

I like to ask a question of large audiences of scientists. I ask if they know the meaning of a certain acronym. The acronym underpins the narrative structure to which almost all scientific journals conform. It is a piece of knowledge that is as central to the lives of scientists as the names on their driver's licenses are to their daily lives.

Speaking to a group of more than 800 scientists at the annual meeting of the American Society of Agronomy, I asked for a show of hands: "Who knows what this acronym means?" I then put up a slide that said simply "IMRAD."

No hands went up. I chuckled, pulled out my cell phone and took a photo of 800 pairs of unraised hands to document the moment for posterity (as well as for any disbelieving scientists, of which I'm sure there are plenty).

Then I asked a second question: "How many of you have ever read a scientific paper that was broken into four sections labeled Introduction (I), Methods (M), Results (R), and (A) Discussion (D)?" By the time I reached the "R," you could hear the chuckles and comments of "Ah, ya got us!"

They have all read hundreds, thousands, tens of thousands of papers that conform to this structure. IMRAD, as I will tell later, was hammered out a century ago and eventually accepted as the standardized structure for how a scientific report is best presented. It is simple in form and essentially identical to the three-act structure that is at the heart of virtually every movie or play written today. It is the structure of a story, which has a beginning (I), middle (M&R), and end (D).

Yet there were no hands raised. And as if to show that it is the exception that proves the rule, it turned out there actually was one

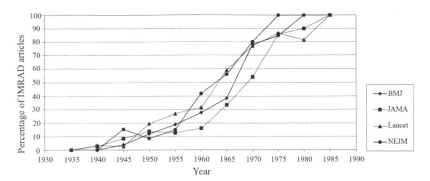

Figure 1. The gradual adoption of the IMRAD Template in biomedicine. We all know today it was the definition of a good idea, and yet . . . look how long it took for the IMRAD form to be adopted completely in the four top medical journals (from Sollaci and Pereira 2004).

hand, way to my left, that was raised, which I noticed only after the second question. Everyone on that side was pointing to him saying, "Here's one!"

I called on him. It was Josh Schimel, author of the popular book *Writing Science Papers: How to Write Papers That Get Cited and Proposals That Get Funded."* He knew the acronym, of course—his book has an entire section on it. But he was the only one.

I did the same stunt with about 200 doctors and students at Johns Hopkins Bayview Medical Center. Same result. Zero hands raised. I've also run the acronym by all of my scientist friends. Nobody has ever heard it, even though there is an entire body of literature around things like the history of IMRAD, the power of IMRAD, the uniformity of IMRAD, and so on. I myself was a scientist for 20 years yet only learned in the past year that there is a formal label for this text structure.

IMRAD: So What?

Okay, big deal, so a bunch of scientists didn't know the acronym that describes the structure of their papers—knowing the IMRAD label is not necessary in order to use it. But what matters is what this reflects. Science is a profession that is permeated with narrative structure and process, yet scientists are so blind to the importance of narrative that they don't even make use of this established label.

If narrative were held up as important, all science courses from the first day would say, "Our profession is so completely built around narrative dynamics, we even force scientists to comply with a narrative template known as IMRAD, which you need to learn about." They might even go on to say, "Narrative and story are pretty much the same thing, which means over a century ago scientists accepted that story is at the heart of their profession. Which means there is no reason for you to have any irrational fear of story." (This last bit might help with the problem of "storyphobia" I discuss in chapter 11.) But none of this happens.

Now you might ask, "So what is at stake if the science world isn't aware of how ubiquitous narrative is and how it works?" The answer is, everything.

Problem 1: Exaggeration Nation

I'm going to use the term "narrative deficiency" to refer to the general problem this book addresses: not enough comprehension of narrative and how it works. Narrative deficiency might not be much of a problem to a plumber or an air traffic controller, but in science, narrative is everywhere. If you don't understand narrative, you don't fully understand science. Let's look at how pervasive it is.

Science consists of two major parts: the doing of science (research using the scientific method) and the dissemination of information about what was done (communication). Both suffer the consequences of narrative deficiency.

On the research side, there are only two outcomes to scientific studies. Either they produce positive results (we saw a pattern), or they produce null results (we didn't see any pattern). The positive result is the same as telling a good story (we saw something!); the null result is equal to telling a boring story (sorry, we didn't see anything, zzz . . .).

The problem these days is that everyone wants to tell good stories while nobody wants to tell boring stories. The journals want to tell good stories, the scientists want to tell good stories, the outreach staff want to tell good stories, and the journalists want to tell good stories. It ends up being a conspiracy of good storytelling. Which can be bad.

In 2014 Petroc Sumner and his colleagues demonstrated the seriousness of this problem for health sciences. They examined biomedical press releases from 20 major UK universities versus the published research papers upon which the releases were based. They found that 40 percent of the press releases contained exaggerated advice, 33 percent contained exaggerated causal claims, and 36 percent contained exaggerated inference.

That's a whole lot of exaggeration, leading to the telling of bigger and more exciting stories than what actually exist in the real world. This is bad news for science, which seeks to document the real world, regardless of how good the story.

This is where I need to be clear on what I'm advocating with this book. It is essential that every scientist understand what makes for a good story. A lot of what I will be presenting will help you achieve that goal. But advocating this understanding is not the same as saying you should necessarily tell only "good" stories.

The problem of good storytelling run amok crops up in the form of what are called "false positives"—seeing a pattern when there isn't one. For example, let's say you announce to the world that ice cream causes cancer when in fact it doesn't (a false positive result). Such a report would probably put you on the front page of newspapers everywhere. People would be excited—the journal in which

you report it, the outreach folks at your university, the journalists who shape your work into a form for the general public—all
revved up. It is enticing and will bolster your career. But what if it's
not true? What if it's a false positive and ice cream doesn't cause
cancer?

In contrast, if your study concludes from the start that ice cream
does not cause cancer, about all you'll get from the newspapers will
be a big "duh."

This sounds silly, but it's the state of the world in science today.
The proliferation of false positives is anywhere from a significant
concern in some fields to out of control in others. Specifically,
the field of biomedical research knows it has serious problems. In
2013 John Ioannidis, MD, of Stanford University, who has become
famous as the chronicler of the current false positive plague in the
biomedical world, announced, "Most of the claimed statistically
significant effects in traditional medical research are false positives
or substantially exaggerated." Notice he didn't say "some." He said
"most."

On a similar note, a prominent geneticist I spoke with recently
said, "Pretty much all the papers published these days in *Science*
and *Nature* in my field are overstated."

Randy Schekman of the University of California, Berkeley, in
his acceptance speech for his 2013 Nobel Prize, even went so far
as to announce his own personal boycott of the top journals, saying that he and members of his laboratory would no longer submit
their papers to the three most important scientific journals, *Science*, *Nature* and *Cell*. He did this because he feels the criteria for
acceptance has been based on "significance" (how big is the story
the paper tells?) rather than "soundness" (how well done is the
research?).

Is the proliferation of false positives a disaster? Probably not. But
it is definitely a significant problem and, according to every scientist I spoke with in the writing of this book, one that is growing.

Furthermore, it has become increasingly clear that, at the same time scientists are reaching for big headlines, scientific journals are less interested in publishing research that doesn't produce big headlines. In 2014 Annie Franco and colleagues published a paper in *Science* titled "Publication Bias in the Social Sciences: Unlocking the File Drawer." They showed how extensive the discrimination against null papers is in at least one field. They found that for the social sciences, null studies had a 40 percent lower chance of being accepted for publication, which in turn translated into a 60 percent lesser chance that the investigators would even bother sending them in to be published—thus their reference to "the file drawer," which is where so many null studies end up languishing.

The bottom line is that positive studies tell big stories and get published; null studies tell small stories and have a hard time getting published. This is all a function of narrative dynamics, meaning how stories are told. It's all the same thing.

One would hope the world of science is accurate, but unfortunately the scientific literature gets pulled away from accuracy by the allure of prominent positive results and discrimination against null results, both of which affect narrative dynamics. (By the way, note the second word in "scientific literature.")

Problem 2: Numbed Down

For the other half of science—the communication of research findings both among scientists and to the general public—the problems are age-old. It's the struggle of connecting with an increasingly numbed populace. Scientists are famous for being bad communicators. I documented this in 2009 with my first book, *Don't Be Such a Scientist: Talking Substance in an Age of Style.* I pointed out the difficulties of communicating information-heavy material. The book was well received, along with several other similar books such as *Am I Making Myself Clear?* by Cornelia Dean, *Escape from the Ivory*

Figure 2. The Narrative Spectrum. Too little narrative content, you're boring; too much, you're confusing. But there's an optimum where you achieve the goal of effective communication.

Tower by Nancy Baron, and *Unscientific America* by Chris Mooney and Sheril Kirshenbaum. The general reception from the scientific community was a collective response of "We know, we're working on it."

The costs of poor communication range from students getting bored with their science classes to the inability of the scientific community to deal with the growing antiscience movements for subjects such as climate science, evolution and vaccination policy.

More analytically, we can look at the communication problems in terms of narrative structure. Eventually I'm going to get into plenty of detail on this, but for now let me just offer it in simple terms. There is an optimum for narrative in communication. There is a certain amount of story complexity that is enough to be engaging, but not so much as to be confusing. It's pretty much that simple.

The same sort of optimum exists for Hollywood movies. Just look at a hugely successful recent movie like *Gravity*. The movie had one main character (Ryan Stone, portrayed by Sandra Bullock), one main incident (the debris cluster damaged her spacecraft), and one clear goal (to get home alive). It didn't have 15 stories happening at once, but just the same, it definitely had one very good story happening. The basic elements were simple, but out of her simple predicament arose all sorts of story complexity. The same dynamic is ideal for both scientific research and science communication.

Figure 2 shows this in terms of a spectrum. Some people don't have enough narrative content in what they are saying and they get boring. Other people are trying to tell you multiple stories at once and you can't follow them, making them confusing. And then there are some people . . . they just have a sense for the right amount of narrative. That intuition in Hollywood is referred to as "story sense." For our purposes, I'm calling it "narrative intuition." Developing this intuition needs to be the ultimate objective for the science world.

I titled the most important chapter of *Don't Be Such a Scientist* "Don't Be Such a Poor Storyteller." I ended the book by pointing in the direction of narrative as the way to address these problems, but I had little specific advice because I had not done that much work on narrative myself.

I followed that book by recruiting two veteran actor friends, Dorie Barton and Brian Palermo, to create a workshop to address this challenge. Over the next four years we taught our Connection Storymaker workshop to a variety of science and environmental organizations, eventually culminating in 2013 with our book *Connection: Hollywood Storytelling Meets Critical Thinking*. The workshops led me to the specific tools and advice I'm presenting here.

My overall conclusion is that the world of science, although steeped in narrative, is largely oblivious to the power and importance of it. This needs to change. And I know who has the knowledge to make the change possible.

Hollywood: Savior of Science?

Right now countless scientists are suppressing their gag reflex after reading this heading. You might be one of them.

In general, Hollywood is anathema to science. Scientists hold the truth as their highest aspiration. Hollywood views the truth as an optional add-on that can be fun if it's convenient. Hollywood's general attitude is reflected by one of the greatest screenwriters of today, Aaron Sorkin. Commenting on his movie *The Social Network*, he seemed to be speaking for the entire industry when he said, "I don't want my fidelity to be to the truth; I want it to be to storytelling."

That is so beautifully put. And I guarantee you it produces everything from shivers to rage in scientists.

Actor/director Ben Affleck put a finer (and even laughable) point on it when he defended his movie *Argo*, which was based on historical events, as having "a spirit of truth." And that's about where you completely lose the entire science community. There is no "spirit of truth." Either something is true or it isn't. Ah, Hollywood.

I share some of this revulsion. I was a scientist. I still have 49 percent of my brain that is programmed like a scientist's. I feel their pain. And yet, there comes a time . . .

Science now needs something that Hollywood has. It's not the ability to make large glitzy action movies that use science to titillate while distorting all the good work of so many humble people. I have little use for those big dumb movies, and I don't think the science world should hold out too much hope for them.

I'm talking about something much deeper. Not the output of Hollywood (what they produce) but rather the process (how they create it). It's the power of narrative. Hollywood is the place that has figured out how narrative works in the real world. Lots of humanities scholars can babble on endlessly about their theories of narrative, but most couldn't spot the basic principles at work in our lives. It's the people in Hollywood who have cracked the code of narrative over the past century, thanks to the driving force of financial profit. Science now needs their help.

Think about it in *Silence of the Lambs* terms. Science, in the form of Agent Clarice Starling, needs to slowly, apprehensively walk down the long, dark basement hallway lined with maximum security prison cells. In the distance is Hollywood, in the form of Dr. Hannibal Lecter, locked up in his cell, glaring insanely through his mask, eyes twitching right and left. Clarice may despise Lecter, but the fact is, she needs his help. He has the knowledge. The time has come to set the prejudices aside—the problems are now far more important than worrying about where the solutions come from.

This is the conclusion I have come to at the end of a 40-year journey. I began my professional life as a scientist. I achieved tenure as a professor of marine biology. But then I changed worlds. I moved to Hollywood, attended film school, worked on movies, made movies, and eventually premiered my movies at the Telluride and Tribeca film festivals, among others.

The journey has led me to focus on the narrative problems of the science world. I firmly believe Hollywood holds the great knowledge that science needs to master. It's time to talk to Hannibal Lecter.

Or if not him, then at least Eric Cartman.

Eric Cartman to the Rescue?

The need to address the problem of narrative deficiency was my big revelation. And who exactly, you might ask, brought me around to this? The answer is simple—Eric Cartman of the animated show *South Park*.

Yes, it's true. Actually, not Cartman himself but his co-creator, Trey Parker. I, like millions of wise Americans, am a devoted fan of *South Park*. So in the fall of 2011, when Comedy Central ran an excellent half-hour documentary about the making of the show, titled *Six Days to Air*, I tuned in.

In the middle of the show there was a scene that changed my life. It was extraordinarily profound, and I believe it can transform the entire world of science. The scene featured Trey Parker talking about his technique for editing the first draft of each show's script. He said,

I sort of always call it the rule of replacing *and*'s with either *but*'s or *therefore*. And so it's always like, this happens, and then this happens, and then this happens—whenever I can go back in the writing and change that to this happens, THEREFORE this happens, BUT this happens—whenever you can exchange your *and*'s with *but*'s or *therefore*'s, it makes for better writing.

His words hit me like a bolt of lightning. So clear. So clean. I had never heard such a simple rule for storytelling. I wrote it down immediately. I've now spent three years researching it, going all the way back to Aristotle (Trey Parker didn't invent the idea). I've given a TEDMED talk on it, published a letter in *Science* about it, and used it nonstop in my workshops.

I've developed it into a simple one-sentence, fill-in-the-blanks template called the ABT (meaning "And, But, Therefore"). The template is this:

_____ and _____, but _____, therefore _____.

Every story can be reduced to this single structure. I can tell you the story of a little girl living on a farm in Kansas AND her life is boring, BUT one day a tornado sweeps her away to the land of Oz, THEREFORE she must undertake a journey to find her way home. That is the ABT at work.

In a more practical way, a scientist could say, for example, "I can tell you that in my laboratory we study physiology AND biochemistry, BUT in recent years we've realized the important questions are at the molecular level, THEREFORE we are now investigating the following molecular questions . . ." That would be the narrative of that particular research program. You can do the same for whatever you are working on.

The ABT is also a tool for creating an "elevator pitch" (a concise explanation of a project) in a way that draws on the power of narrative structure. We will get into this in great detail in part 3.

The Hegelian Way

"The ABT is the DNA of story." That is what Park Howell, a professor who teaches storytelling in the business school at Arizona State University wrote to me recently. I believe this is correct and is not an exaggeration. The ABT really is that powerful and profound.

But then guess what I discovered as soon as I started talking in terms of DNA. Two other authors think they've discovered the DNA of a similar skill—argumentation—in the form of their own template.

In the enormously popular textbook *They Say, I Say* (it has sold over a million copies since publication in 2006), Gerald Graff and Cathy Birkenstein help you find the structure of your argument using templates. They start with the simple idea of presenting what your opponents say first, then what you have to say, before reconciling the two.

At the start of their book they say, "The central rhetorical move that we focus on in this book is the 'they say/I say' template that represents the deep, underlying structure, the internal DNA as it were, of all effective argumentation."

There you have it—two skills—storytelling and argumentation. Traditionally they have been seen as polar opposites—one has fun with the truth, the other tries to find the truth. And yet, there is a similarity of structure.

Look at the two templates—the ABT and "they say, I say." See any similarities? Both begin with the setup (a few facts in the ABT, what others say for argumentation), then establish a problem (using "but" in the ABT, telling what *I* have to say for argumentation), then the resolution of the two parts.

It's no coincidence that the templates are so similar. They are derived from what is really the true DNA of just about all interesting thought. It's called the Hegelian Triad or the Hegelian Dialectic. It was first identified by Georg Hegel, the great philosopher of the late 1700s and early 1800s. It has three parts—thesis, antithesis, synthesis—just like these templates. It underpins pretty much every-thing from logic to reasoning to argumentation to storytelling. And guess what it also underpins—the scientific method.

So there's your real DNA. The Hegelian Triad is so powerful and universal that I've broken this book into the same three elements. My concern is the need for more awareness of narrative in the science world. I begin the book with "Thesis," where I describe the state of the science world today with its deficient awareness of nar-rative despite the ubiquity of narrative within it. Then I present "Antithesis," where I lay out a set of tools that could remedy this problem yet are not widely in use. Finally I pull it all together with "Synthesis," where I tell of the effectiveness of the tools and offer up my prescription of Story Circles as a means of propagating this knowledge.

I've also used another element of structuring. The first and third sections (Thesis and Synthesis) follow the ABT Template. Together these give the book narrative structure at multiple levels—just like a fractal pattern, which repeats itself at all scales (more to come on this). In fact, I like to say that the three letters should also stand for "Always Be Telling stories." We'll get much deeper into that, but let me now say the same thing in a different, perhaps more shocking, way . . .

Science Needs to Emulate Trey Parker

Now you're thinking I've totally lost my mind. I'm recommending the entire science world become more like Trey Parker. How in the world can I be saying this? Have my years of living and working in Hollywood made me into one of the lunatics in the asylum? Possi-bly. But first, hear me out.

In *Don't Be Such a Scientist* I looked at knowledge in terms of the cerebral versus the visceral. Academia with all its information is the place for the cerebral, while Hollywood—the land of emo-

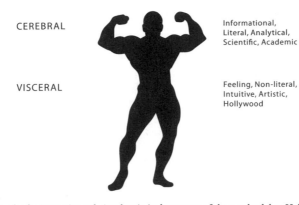

CEREBRAL

Informational,
Literal, Analytical,
Scientific, Academic

VISCERAL

Feeling, Non-literal,
Intuitive, Artistic,
Hollywood

Figure 3. Cerebral versus visceral. Academia is the master of the cerebral, but Hollywood wins when it comes to the visceral. For success with narrative, you need both.

tion—is more about the visceral. College professors are masters of the cerebral but not that great when it comes to the visceral. Hollywood is the opposite—populated by plenty of bean brains but owning the visceral end of the spectrum, able to arouse the masses like no other force.

So let's look at Trey Parker. When it comes to storytelling, he is a 500-pound visceral gorilla. He has no graduate degrees. He's no scholar (and I'm sure he would be the first to admit this). He got his undergraduate degree at the University of Colorado then moved to the "storytelling gym" (Hollywood) and began lifting storytelling weights nonstop, day in and day out.

He put the burn on his storytelling biceps, starting in 1997, by telling stories, week after week, which *had* to work. He wasn't given the basic academic luxury of living a life of three options (yes, no or later). He and Matt Stone were put into the pressure cooker of storytelling with their animated series *South Park*—either figure out how to tell stories that work, or fail and go home. Do or die.

By 2011, when they shot the Comedy Central documentary, Parker had become narratively muscle-bound. *South Park* was by then the greatest hit in the history of Comedy Central, and Parker and Stone's musical, *The Book of Mormon*, had taken Broadway by storm, winning nine Tony Awards. Parker's brain had become buff with narrative muscle. With that strength he was able to distill much of the whole story development process down to his simple rule of replacing *and*'s with *but*'s and *therefore*'s (which, as I will explain later, he picked up in college and which probably originated with one of the greatest screenwriting instructors of all time).

The Goals: Narrative Intuition and Narrative Culture

Trey Parker and many of my USC film school classmates possess what scientists need—narrative intuition. Narrative intuition is the ability not just to know the basic rules of narrative but to have absorbed and assimilated them so thoroughly you can actually sense them. In essence, to be like Trey Parker.

I have seen narrative intuition in action over the years with veteran screenwriters. They have two abilities. First, they can create stories that are concise and compelling, and second, they can listen to stories that are *not* concise and compelling and quickly figure out how to fix them. They have an ability to hear a story and immediately pinpoint why the story is boring or confusing.

If scientists had this trait at a deep level, it would enable them to fix or avoid many if not most of the problems I identified earlier. They would be more sensitized to the dark sides of storytelling. They would be less inclined to unknowingly make the mistake of false positives. If they understood and prioritized narrative, they would reduce the publication bias against null results. And if they had an intuitive feel for narrative, they would write and speak in a manner that was less boring, and not as frequently confusing. It is the change that is needed for the entire profession. No, narrative intuition is not a panacea (always, the science-minded will set to work picking holes in any proposition by taking it to the extreme), but it is a means of addressing a source of many problems.

Narrative is incredibly powerful, not just as a tool for the workplace but for making sense of the world. My purpose for writing this book is to urge scientists to put narrative on the score sheet. It should be one, if not the highest, priority for all science programs and agendas.

Achieving this intuition in a profession that is so steeped in narrative is the only long-term hope for combatting the problems facing scientific research and science communication. Instruction in narrative dynamics needs to happen at all levels, and especially at the very beginning of science education, so that recognizing and creating narrative can become intuitive.

If multiple individuals within an organization achieve narrative intuition, a "narrative culture" can develop. This culture can establish expectations and standards for a minimum level of narrative quality. Norms can change when everyone is expected to have a certain level of familiarity and competence with narrative dynamics. Once this happens, the secondary effect of "entrainment"—where people are swept along with the flow—can occur, making the new norms self-perpetuating.

This is not an unrealistic hope. The necessary tools are in this book. It's just a matter of embarking on the mission to make it happen, so let's get going.

The journey starts now.

II

THESIS

Aren't We Special

Ah, email. The crazy thing about it is there's no inflection. At least with handwritten communication the writer can bear down, draw the letters erratically, and add underscores and scribbles to convey rage or affection. But with email, there's nothing but electronic letters and maybe a few annoying emoticons. The result is often the worst possible interpretation.

Such was the case as I headed out the door for my "cooling-off jog" after receiving the shocking email that opened with, "Well, Randy . . . aren't we special." I read the worst possible sentiments into it. Actually, it was kind of hard not to, given the buildup.

The next morning I called Megan, poured out a string of heartfelt apologies, then said I was withdrawing from the event. I told her how much respect I have for the two scientists and how I never wanted to get into such a war of words. She not only accepted my withdrawal, she overrode my apologies with her own, saying she had no idea how personal this stuff could get. She originally thought they would be thrilled to have my assistance. It never dawned on her it would turn so hurtful, but now that things had gone wrong she understood my decision.

We reached the end of our apologies, she accepted my withdrawal, and I was in that final singsong stage you enter as a phone call is wrapping up. But then an email appeared in my inbox from one of the two scientists. I read it aloud to her.

He and the other scientist had talked the night before. He said, "We decided we're senior enough and we've given enough successful presentations that we can afford one complete debacle, so we're gonna go ahead and roll with your crazy ideas—nothing ventured, nothing gained."

Well . . . . aren't we really special. Deep inside I knew these two guys were amazing and that's why I had offered to withdraw—you don't want to end up in disagreements with people you have such respect for. So I was instantly psyched. I uncanceled things with Megan and within minutes was on a Skype chat with him.

Houston, we had lift-off. Whew.

And now we're ready for lift-off in our journey into the world of narrative, so let me begin by introducing your host, me, in a little more depth. I have a unique background that is central to my message. I spent half my career as a scientist, the other half as a filmmaker. I am bilingual in the languages of "academic science" and "working Hollywood." So here I am.

Randysseus

People like to have fun with my first name, calling me everything from Rrrrranders to Randymon, Randosius, Randcho, Randango, Randitola, the Rand Man, and . . . you get the idea. I think it's a pretty dumb name, more fitting of the Randy character on *South Park*, but what can you do?

So now I am presenting my own version of my name, which is Randysseus. I offer this name because I, much like the legendary figure Odysseus (well, okay, not that much, in fact hardly at all, but roll with me on this one), have been a sojourner of great distances. At least psychologically.

I lived my early life as a scientist. I earned my PhD in biology at Harvard, spent a year living on an island on the Great Barrier Reef of Australia, went diving under the ice in Antarctica, dove a half-mile down into the deep sea, lived in an undersea habitat at a depth of 60 feet for a week, and did pretty much every other exciting and interesting thing in the ocean I ever yearned to do.

Eventually I became a professor of marine biology, I had graduate students, scored major research grants including a significant one from the National Science Foundation, published 20 peer-reviewed research papers including one in *Nature*, and finally was awarded tenure at the University of New Hampshire. All of which meant I had succeeded as a scientist and was set for life. I had hit that point most academics dream of where you have achieved a "guaranteed job" no matter what happens to you (short of major felony charges—about the only thing that will get you fired from a tenured position).

BUT then . . . (now the actual story begins—a key point of structure, as we'll discuss in detail later) I departed from the comfort of my Ordinary World (another term to come) and set out on a journey (which is what a story is—ah, so much for us to cover!).

From the Land of Science in the East, I headed west to California. It was 1994. Things were primitive in America—Friendster, Auto-Tune and Crocs had yet to be invented. As much as I loved doing science, I had developed an even greater interest in the communication of science. I resigned from my professorship with a big vision in mind. I fully intended to return one day to the world of science and share what I had learned.

I let my friends have fun viewing this as an impulsive move. Rumors went around of my having had an identity crisis or a psychological meltdown. A friend sent me a cartoon of two middle-aged men at a bus stop, one dressed as a pirate, the other dressed as a cowboy and saying, "Midlife crisis?" as the first nods in agreement.

Figure 4. Hollywood agent cyclops.

But my closest friends knew I had a clear purpose and would return some day.

Feeling like Odysseus, I set out for the Sea of Hollywood, ready to confront the Cyclops agents and lawyers, hoping to avoid being seduced by the Lotus Eaters of Malibu and of course determined to steer clear of the Sirens that lurk at every nightclub and Hollywood party.

Just like Odysseus, I pulled it off—nobody destroyed me. I survived it all and I, Randysseus, am now returning to the science world—back from my 20-year journey (twice as long as Odysseus!)

and ready to share the knowledge gained. And here's one of my first realizations in looking at the world of science from outside . . .

Scientists Love Complexity

Story dynamics, which thrive on simplicity, don't mesh well with science, where complexity is the norm. I know this well from my cross-cultural journey. I have experienced the challenges firsthand.

In *Don't Be Such a Scientist*, I shared a number of classic moments where my "scienceness" stuck out in Hollywood in comically embarrassing ways. Here's one I left out—yet another tale from my early days in Hollywood, the former professor of marine biology making his way in Tinseltown. This particular story brings to life the complexity/simplicity divide.

At USC film school I was chosen for one of the four director positions in my class. I was given the equivalent of a $50,000 budget to make a short film from a script I had written for a wild and outlandish musical comedy. It was the story of a woman in law school who gets electrocuted in her kitchen while angrily making dinner for her husband and his business partners the night before her big law exam. Her ghost comes back for revenge in the final scene, teaming up with the secretaries from her husband's office to perform a song-and-dance number about castration. Suffice it to say, the film was a little out of the ordinary in the refined atmosphere of the USC School of Cinematic Arts. (I was eventually accused of being a misogynist, even though the guy got the punishment—such are the politics of film school.)

For the dance scenes I managed to recruit a wonderful choreographer named Lance MacDonald who, the next year, became the assistant choreographer on a little movie called *Titanic*. On our first day working together, I showed him a set of diagrams of the dance scenes I had diligently created.

They looked like football plays as I laid them out on a table for Lance to examine. The diagrams were filled with X's and O's for the dancers, with arrows for who should go where at each moment, then a bunch of V's that marked where the camera should be placed at different points along the way. The diagrams looked so precisely drawn, and yet . . . I didn't know the first thing about choreography, and it had only been a year and a half since I was a scientist laboring away in a laboratory. The diagrams were the picture of an analytical mind at work.

Lance studied them for a while in fascination. Then he got up, picked up the diagrams and walked to the other side of the room saying, "Wow, these are really, really great—I'm so impressed with the work you've done," as he dropped them into a trash can. "But we won't be needing them—you'll see." I was stunned, but with time I came to realize what he meant.

Stupid me. Stupid, stupid me. The scientist, mired in his complex, complicated charts, thinking I could plan out every detail. I would come to learn that choreography is about art, and at the core of art is simplicity.

Lance set to work, hiring the dancers, beginning rehearsals in a studio, and then after a couple weeks he invited the cameraman to a session for the next step. He ran the dancers through their moves. He had them freeze at various moments while he and the cameraman looked at them from all angles and figured out where they would put the camera. Instead of having them dance for the camera, he had them dance for themselves while the camera documented it. And I just sat and observed—the misguided novice.

Had I insisted on forcing my diagrams on the production, we would have had a clunky mess of dancers bumping into each other and trying to find their way to key points at exact times to line up with the camera angles. Instead Lance produced a smooth, flowing, organic, fun performance that ended up being shot flawlessly. So simple. So perfect. So sophisticated.

It was the embodiment of a famous quote usually attributed to Da Vinci . . .

"Simplicity is the ultimate sophistication."

I tell this little tale because it strikes at the heart of this book. It also gets at what I think is the problem with a lot of today's books and workshops addressing this suddenly ubiquitous topic of story and storytelling. Most of them are as mired in complexity as I was with my diagrams.

I see books that are packed with charts and graphs and section after section about protagonists, antagonists, sequences, culminations, narrative arcs, tropes, themes and . . . it's all so exciting and stimulating with the complexity, but in the end, is it effective and necessary?

That's the problem with complexity. It can be so overstimulating that the net result is zero, as the recipient can't ever lock onto one thing to retain. It's like standing on a cliff above a city, taking in the stunning view. It may be magnificent, but you will probably walk away with little to say other than, "Wow, that was awesome!" You saw it all but retained virtually nothing.

In contrast, I take a sort of fractal approach to the entire concept of story. At the core of fractal design is the basic idea that "out of simplicity can arise complexity." It's just like an ice crystal, which looks amazingly complicated at first glance, but on closer inspection is just a single pattern that has been replicated over and over. A pattern like the ABT.

I firmly believe this is the case for all storytelling, and I'm not alone. John Yorke, in his wonderful book *Into the Woods: How Stories Work and Why We Tell Them*, addresses the fractal nature of stories in detail. He writes, "Stories are built from acts, acts are built from scenes and scenes are built from even smaller units called beats. All these units are constructed in three parts: fractal versions of the three-act whole." Out of that singular, simple structure can arise endless complexity.

Complexity is fun, exciting, and stimulating and can be non-repetitive, which is a central element of entertainment. But I am advocating for simplicity—just a few basic tools, used over and over again as you develop an intuitive feel for how narrative works. If your attention span is so short that you can't bear the thought of repeated actions, then I'm not sure you're ever going to develop much of a feel for narrative. Narrative is about simplicity and repetition.

I will come back to this theme repeatedly. Scientists like to attack things for being "dumbed down," which is at times a fair criticism, but you can't afford to let "dumbed down" get confused with simplicity. There is a difference. As Da Vinci pointed out, simplicity is about sophistication.

Now let's begin considering the world of science, and the larger world of narrative into which it arose.

1

Science is stuck living in
a narrative world . . .

The Long History of Narrative: Gilgawho?

Raise your hand if you know who Gilgamesh was. This is another demonstration I do with groups of scientists, and again, I'm no better than any of them. I had no clue who Gilgamesh was until a few months ago when I began writing this book—such are the holes in my humanities education.

Gilgamesh is pretty much where the entire concept of narrative begins. His story is humanity's first story, and is the birth of literature. He was a great and mighty leader 4,000 years ago who supposedly ruled Mesopotamia for 126 years. Early storytellers carved his epic tale on stone tablets. After Gilgamesh, the rest is literally history as we became a "storytelling animal," which happens to be the title of a nice book by Jonathan Gottschall published in 2013. Gottschall makes the point that story (or narrative if you prefer) pervades every aspect of our lives today.

From Gilgamesh we jump 2,000 years to the next major story milestone, namely, Aristotle and the Greeks. Aristotle realized that stories have a distinct structure. In the *Poetics* he talked about the structure of plays and stories. He broke them into five basic parts. The opening he called *prologue*, the ending he called *exodus*,

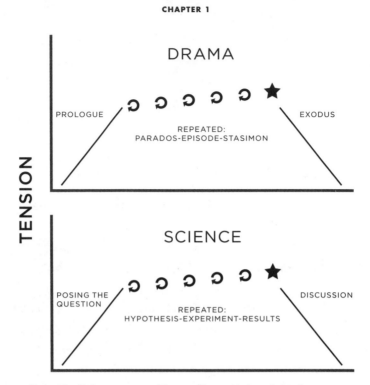

Figure 5. Dude, it's all the same story. The top diagram is how Aristotle, 2,000 years ago, described the structure of a story. The bottom diagram is how a scientist conducts a research project. See any similarity?

and in the middle he described a series of repeating cycles, each consisting of three parts—*parados, episode* and *stasimon*. Even today, when we think of a story, we often talk about the middle part being "episodic."

Now here is the major revelation in thinking about storytelling in science. What are the major parts of a scientific project? You start by gathering background knowledge (introduction), then you repeat cycles of posing and testing hypotheses (methods and results), until finally you discover an answer. At that point you pull it all together with a discussion.

Take a look at those two structures side by side in figure 5. This is the first example of the point I will hit on repeatedly: "Dude, it's all the same story"—which is what my *Connection* coauthor Dorie

Barton began saying to me in our workshops as we discussed story structure. Initially the scientist in me bristled at this suggestion— surely there are many, many types of stories. But these days I find myself pretty much in agreement with her and hope you will as well. This stuff really does come down to the same core structure.

The structure that underpins a story, of course, radiated into all sorts of variations over time. There are so many seemingly different types of stories—romance, horror, comedy, fantasy and more. If you choose to do so you can get lost in the infinite complexity of them. Similarly, you can get lost in the infinite complexity of biological diversity—marveling at everything from the bizarre shape of a guitarfish to the roiling, hydra-like living ball of spaghetti that is a basket starfish. You can sit there and marvel, saying, "Wow, each creature is so different from the next—there is so much complexity in the various species of life!"

And yet . . . dude, at the core, their DNA is telling the same story. Their genomes track back to the same original primal sequences of base pairs. You can choose to focus on the mesmerizing complexity, or you can seek the simplicity at the center of it all. Finding the simple core allows you to say, "I see how all these various forms branched off from the one original type." The former is exciting but ultimately directionless. The latter makes sense of the world.

It's the same with stories—all variations track back to a common heritage. This is what anthropologist Joseph Campbell realized in the first half of the last century. He brought an analytical perspective—essentially the mind of a scientist—to the traditionally nonanalytical world of storytelling. Just as an evolutionary biologist looks for common descent among organisms, he looked for common structure among stories told by different cultures and religions around the world.

In 1949 he wrote his landmark book *The Hero with a Thousand Faces*, which he opened by saying, "There are of course differences between the numerous mythologies and religions of mankind, but this is a book about the similarities." That was the prelude to his eventual message: "Dude, it's all the same story."

Campbell saw a single, common structure underlying storytelling around the world. He named this structure the *monomyth*. And guess how many major parts there are to it—three. Beginning, middle, end. Just as Hegel would have predicted.

Speaking of the number three, guess what else emerged over the centuries in the world of storytelling—the basic structure of plays, novels and eventually movies known as the three-act structure. Today it is at the core of just about every movie you watch—the same tripartite structure, deeply embedded in the programming of the brain. So deep that you can't escape it.

The Short History of Science

Now it's time to think about the history of science. If we know humans have been telling written stories for at least 4,000 years, how long have we been writing scientific papers? The answer is less than 10 percent of that—or about 400 years.

There were scientists long ago. There was Ptolemy in Egypt just 100 years after the time of Christ and the amazing Ibn al-Haytham (nicknamed Ptolemy the Second) in Mesopotamia who, almost 1,000 years later, pioneered optics and experimental physics. But the reporting of scientific research in formal journals began in 1665 with the first volume of *Philosophical Transactions of the Royal Society*. The early reports of scientific research were written very much in the "literary" style—not broken into sections, but more of a single essay, often written in the third person, along the lines of "Recently Robert Boyle conducted a demonstration in which he . . ."

Within 50 years the articles moved from largely descriptive pieces to a form that was beginning to take the shape of reports of experiments written by the investigator. By the late 1800s a clear structure had emerged around the pattern of theory, experi-

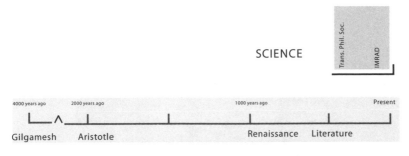

Figure 6. Science is late to the party. The humanities are at least 4,000 years old. Science is a much more recent arrival.

ment, discussion (hmm . . . three parts, what a coincidence). This eventually gave way in the 1900s to the almost universally agreed upon scientific paper template of today that I mentioned back at the start—the mighty IMRAD. (Please tell me you will never forget that term from here on.) And though there are four sections to the IMRAD, the Methods and Results sections are often combined, reflecting the fact they are, after all, just the middle of the story.

Science is a newly arrived guest in an ancient narrative world. This is the challenge scientists face. Scientists might dream of communicating in a nonnarrative form, where all you do is list information, but ultimately that doesn't work. As I pointed out with the "see it, say it" conundrum, it just isn't that easy. Let me take this a little deeper.

The Programming of the Brain Is Defective

People can listen to a few facts, but not many. After a while their narrative need kicks in. You can give a lecture that is pure information with no narrative structure, and a nontechnical audience might be able to endure a half-hour or so before walking out, but that same audience will listen to hours and hours of good stories.

Figure 7. The human brain. If it were for sale it would have been recalled by now.

You could show them a *Breaking Bad* marathon, and they would have no trouble sitting through lots of episodes. Such is the power of narrative.

So that's your first "faulty programming" aspect of the brain. Scientists wish they could just pour out facts, untouched, for consumption. But they can't. The brain needs information packaged in specific ways. This leads to all sorts of distortions that can confound even the best intentions. *New York Times* columnist Nicholas Kristof points this out quite clearly in a classic article that I encourage my workshop participants to read over and over again.

Nicholas Kristof Warns the World about Storytelling

Nicholas Kristof is a two-time Pulitzer Prize winner who wrote an amazingly short, simple (my favorite attribute!) and broadly practical article about the power of story in mass communication. I'm not sure he would approve of my calling his article an essay full of warnings about storytelling, but that's pretty much what it is.

The somewhat surprising thing about the article is that he published it, not in an academic journal or as a *New York Times* feature, but in the November 2009 issue of *Outside*. The title, very fittingly, is "Nicholas Kristof's Advice for Saving the World." If I were to summarize the message of the article in one sentence, I would say, "You should realize how faulty the programming of the human brain is before you set about trying to alter other people's brains."

His point is that communication is not about telling people what you think they need to hear or know; it's about figuring out your goals then working backward, mindful of how the brain works, to successfully convey your message. You need to shape your information into the right form for it to work properly when it enters people's brains.

Another way to say this is by dismissing the Golden Rule. I was raised in Kansas with the charge "Do unto others as you would have them do unto you." Kristof's point is, who cares how you want people to do unto you? What you need to figure out is how people want to be done unto, then work within those constraints.

This is a fundamental problem scientists run into. When they can't figure out why people aren't interested in what they have to say, they get frustrated. They say things like "people need to know this" as they talk about things like the need for greater "science literacy." Of course I agree with their intentions, but before you get angry at "people," you really need to have a basic understanding of how people think.

Singularity: The Power of Storytelling Rests in the Specifics

The most important dynamic Kristof presents, in all its frustrating injustice, is the "power of one" in storytelling, or what we could call the power of the singular narrative. Here's basically (in my paraphrasing) what he says: If I tell you the story of one little girl in Africa who is going to die next year from a disease, you are going to get upset by exactly X number of "units of upset." But if I tell you the story of two little girls in Africa who will die from the disease next year, wouldn't you think you would get twice as upset?

That's twice as many people who are going to die. Think of all the pain you will feel for the first girl's family. Then think about that same pain for the second girl's family. One plus one should equal two, right? It makes sense that you would get twice as upset. But you already know that's not the case. How could it be? You would run out of upset units pretty quickly as the number of victims grows.

This is the sad, illogical, counterintuitive, even dangerous nature of storytelling. Kristof points to the famous saying that "the death of one individual is a tragedy, the death of a million is a statistic." Therein lies the frustration for scientists—how can it not be as simple as just numbers? A million is so much greater a number than one. Isn't it the same as "sample size," which you always want to maximize?

I'm sorry, but it just isn't about the numbers alone. People care about things that move them, touch them, reach inside them, connect with them—all of those things. The story of one person can do all that to you, very powerfully. But it's harder for five people to do it to you, really hard for one hundred, and for a million . . . the people just become a statistic from which you are detached. Which is kind of like what I was saying about taking in the view of a city—so much, and yet so little that will last.

This is a core principle of narrative that you need to commit to heart, and even if you do, you'll still probably make mistakes with it at times. If you don't grasp it, you will be one of those speakers who talks about the 18 different things going on in your lab, all of which you feel passionate about but none of which end up making enough impact for anyone to remember the next day.

It's the "less is more" thing. And it's really, really hard for scientists to grasp. How do I know? Because I used to be a scientist, and

I gave the sort of talks that involved 73 slides in 12 minutes. And in fact I still do it sometimes—what can I say, the wiring of my brain is faulty. But at least I've developed a little bit of awareness. So I present this just to make the initial point that narrative dynamics can be fickle. And dangerous. And underlying this is one of the most important rules in narrative, which is that the power of storytelling rests in the specifics.

A story that lacks specifics is not powerful. Politicians often give boring speeches because they don't want to get locked into specifics if they get elected. They say, "If you elect me I will improve our community." The crowd asks how. The politician replies, "In all the ways it needs improvement." The crowd gets bored. They need specifics to stay interested.

If you think about this phenomenon, you see how it applies here. The quantity of one is as specific as things get. Two is less specific. One is where the power is at a maximum. And guess what this rule also reflects—simplicity. The story of one person is simpler than the story of two people.

"People like a simple story." You hear this refrain all the time and it's true in many ways. It's very frustrating to people who want to communicate the truth, which can be complex, but it's what works. And that becomes the challenge—communicating complicated things in simple ways.

Think of what this means. If you go to Africa and get to know three little girls in a village who are dying from a disease and you want to motivate people in America to donate money to save them, your first instinct might be to tell the more complicated story of all three little girls. It's only natural that you'll want to be "inclusive" and mention all three in equal measure. But the sad truth is, if you really want to help them all, you should pick one and tell her singular story in as much depth, power and detail as possible. You will have a higher chance of actually motivating people. All three will benefit the most by your making that decision, as difficult as it might be.

Please note, I am not advocating that you present the research project you conducted with three colleagues as just your story alone. Sometimes it doesn't matter that the story of just one person is more compelling to the broad audience—you don't want to cheat your colleagues. The only thing I am saying is, it's crucial that you understand these fundamental narrative dynamics and use them to your advantage when appropriate.

Telling a simple story can be frustrating, but it may be the single most important challenge for all scientists. The tendency of scientists to present endless piles of facts, unable to find the singular narrative on which everyone can focus, has been a reason many important science stories, including that of global warming, fail to resonate with the public.

Why So Keen on Narrative?

So what's the big deal about narrative? Why is everyone talking about it? Let me address the power of narrative more scientifically by looking at how it works and why it is so beneficial. Just as one small but robust example, let's take a look at a neurophysiology project addressing the effects of narrative.

In 2008, Uri Hasson and colleagues developed the field of neurocinematics. They used functional magnetic resonance imaging (fMRI) to examine the brain activity of people viewing film clips with and without narrative structure.

Now let me make clear, I am skeptical about today's popular neurophysiology stories. I loved Adam Gopnik's 2013 New Yorker article, "Mindless: The New Neuro-skeptics," and I'm a fan of the British blog Neurobollocks: Debunking Pseudo-Neuroscience So You Don't Have To, which hits the same notes of skepticism. In light of this, I was impressed that when I spoke with Hasson about his work, he was quick to emphasize its limitations. I tried to ask whether they had measured brain response to all kinds of tiny subtleties. He shuddered and warned me repeatedly about the limitations of fMRI. What I'm presenting of his work here is, as you will see, pretty simple in terms of interpretation.

Hasson's group looked at narrative structure as being the same sort of continuum I laid out in the introduction. At one end of the continuum are clips from suspense films by Alfred Hitchcock (highly narrative). At the other end are clips of people walking idly around Washington Square Park (nonnarrative). The fMRI reveals two main things:

1. NARRATIVE ACTIVATES THE BRAIN. People watching narrative clips had much greater overall brain activity than those watching nonnarrative clips.

2. NARRATIVE UNIFIES THE THINKING OF A GROUP. Yes, I know "group think" can be a bad thing, but this effect is not the same as group think. When a group of individuals are being told the same story and it has a strong narrative, their brains will show similar patterns of activity. When people watch clips with strong narrative (Hitchcock), the pattern of brain activity from one individual to the next is much more similar than when there is little narrative (Washington Square Park). Hasson's group calculated an index of similarity (termed Inter-Subject Correlation, or ISC) across subjects and found an ISC of 70 percent for narrative results versus 10–20 percent for nonnarrative.

None of this is particularly surprising. All you have to do is look at the audience during a suspenseful movie scene—say, a man has a gun pointed at him. Pretty much everyone is thinking the same thought: "Is he going to get shot?" In contrast, if you show people

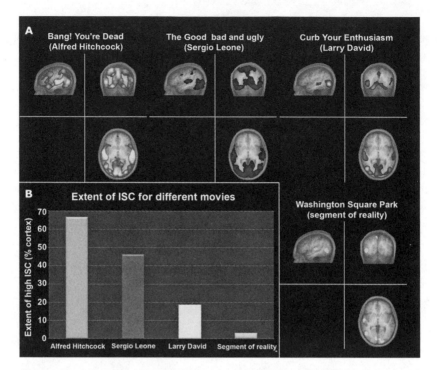

Figure 8. Neurocinematics. Using fMRI, Hasson and colleagues record the brain activity of subjects viewing film clips that have strong narrative structure (such as a suspenseful scene from an Alfred Hitchcock film) and no narrative structure (people wandering aimlessly in Washington Square Park). Their index of similarity in brain activity across viewers (termed Inter-Subject Correlation, or ISC) shows that strong narrative content results in much greater similarity in brain activity across individuals. Used with permission.

nonnarrative clips, their minds will start to wander. What this means is that everyone walks away from the Hitchcock scene thinking the guy is going to get shot, but people viewing the Washington Square Park footage will have much more varied experiences—some saying the clip was about pigeons, others saying it was about people sitting on park benches, and so on.

This then becomes a significant distinction between when and where you use the power of narrative in the writing of a scientific paper. Science advances through the hypothetico-deductive approach. Once the data are in and you have presented them to us, your audience, for our own interpretation, we do indeed look to you to draw on all your experience and knowledge to tell us what you think it all means. That's the entire idea of "discussion."

For effective communication, what you want is focus—everyone thinking the same thoughts and presumably using large amounts of their brains. (We're going to assume the amount of the brain that is active is proportional to the quality of thinking.) Such is the power of narrative: it enables you to pull everyone together. And that, of course, is what clear and obvious problems do—they unify people in their thinking.

The United States of Problem Solvers

How unifying is narrative? Just look at World War II—perhaps the greatest unifying exercise in the history of humanity. Everyone in the United States faced the same problem—how to win World War II. Everyone joined together with every ounce of their strength and eventually succeeded. Most of those people ended up with one thing in common after the war—it turned out to be the most important experience of their lives.

This is the subject of the classic book *The Good War*, by Studs Terkel, winner of the 1985 Pulitzer Prize for nonfiction. Terkel tells all about how WWII was the most meaningful experience in the lives of an entire generation. Even people who never came close to the war—who spent the war working in a bakery—during those years felt their job baking bread was important as part of the overall war effort. Once the problem was solved (the war was won), most of them said their lives never again felt as meaningful. (Note: keep this term *meaningful* in mind for later when we get to the Dobzhansky Template.)

Recognizing the problem-solution dynamic at the core of narrative is the starting point for many people who come to me with stories they can't make work. I begin by just asking, "What's the problem at the core of this material?" And they reply, "Well, we're wanting to tell the public about our wetlands management program." And I say, "Right, so what's the *problem* at the center of your program?" And they reply, "We've destroyed too much of our wetlands." And I say, "There you go, we're starting to get somewhere. You've identified a central problem that you're trying to solve: How do we stop destroying so much of our wetlands? Now you can tell a story around that problem."

Storytelling is about identifying the problem being addressed. It's about knowing Scarlett O'Hara's problem at Tara, Rick Blaine's problem in Casablanca, E.T.'s problem on Earth, Luke Skywalker's problem in a galaxy far, far away. On and on and on. It's all about problems and solutions, which is what science is also about, so you'd think scientists would be better with narrative.

Why aren't they?

Sprinting Past the Humanities

As a scientist I never quite knew what I had missed in my education, but when I left the comfort of the Biology Department at the University of New Hampshire and moved to the intensely narrative world of Hollywood, it became clear what had happened.

I thought I knew so much as a tenured professor. I fully expected Hollywood to be challenging, but I wasn't aware of how much my sprint to become a scientist had left me ill-equipped to deal with other aspects of life. It took only a week of film school for the shortcomings to become obvious. Almost all of the 50 students in my entering class had undergraduate degrees from a humanities discipline—English, history, art, music. I was the only scientist. They understood narrative at a fairly intuitive level; I didn't.

This was a handicap that showed in my musical comedy short film, despite the awards I received for it. It was a wacky, wild film that audiences enjoyed, but it didn't tell much of a story. And yet despite this hole in my background, the overall experience of entering the film world from science wasn't all that alien. The characters and setting definitely were different, but there was a similar feel to the process. So many people ask me about how weird it must have been to go from science to cinema. I always reply that I have been much more impressed by the similarities in the two worlds than by the differences.

So, at the start of my journey, at age 38, I knew not the first thing about stories or how important and pervasive they are. Why should I have known anything? I was trained as a scientist.

And there you have it. The basic problem. Scientists are trained only as scientists—meaning they sprint past the humanities to the best of their ability when they are undergraduates. I certainly did this, and I know I'm not alone.

A couple years ago I was talking about this with a group at the American Association of the Advancement of Science—the world's largest science organization. The head of AAAS at the time, Alan Leshner, broke in to say, "Same with me—when I got to college, all I wanted to study was science, so I bypassed almost all of the humanities courses."

Is that just a thing of the past? Hardly. I spoke with Stephanie Yin, who worked as one of my assistants right after she graduated from Brown University. At Brown they let you design your own curriculum (no surprise for the uber-progressive Brown). Guess what she did—same thing. She arrived at college, knew she wanted to be a scientist, and went straight for the science courses. She took a creative nonfiction class, a graphic novels course, a history seminar on Chinese Americans and a few semesters of Hindi. That was it. No History of American Literature, no Western Civilization, no Shakespeare—pretty much nothing that didn't have science in it.

At one of my workshops with postdocs of National Institutes of Health I asked a more important question. How many of them had received some sort of training in basic narrative principles somewhere along the way in their educations? The answer was none.

Of course, scientists are trained to be critical and pick holes in everything you say, so at this point I'm sure many readers are thinking, "That's not a data set, and I have lots of scientist friends who took plenty of courses in the humanities." Yes, I do too. But trust me, overall, the majority of science students bypass the humanities and I don't blame them. Science is fun. Why waste time reading Dickens and Chaucer when you can search for the definition of life itself?

So why does this matter? Let me start by mentioning a few famous scientists you'll hear more about later in this book. First off, James Watson. Not only did he co-discover the structure of DNA, he also wrote a magnificent book about his experiences that has stood the test of time. We will break down the narrative structure of that book in part 3. (And yes, I am fully aware of the claims against Watson for failing to cite people who deserved shared credit in this discovery. But think about that in the context of what I said about the power of singular narrative. He knew it well and used it to his own advantage as a communicator. No, I am not suggesting that you emulate Watson. He used these principles in the wrong way, to the detriment of others. But I am advocating that you, like Watson, develop a deep understanding of how narrative works and use it to your advantage *when appropriate*.)

Guess what Watson's undergraduate education was like. In his autobiography, with the delightfully curmudgeonly title *Avoid Boring People: Lessons from a Life in Science*, he tells of his greatest teachers as an undergraduate, saying, "Particularly moving was Green's Humanities II lecture on the grand inquisitor of Dostoevsky's *Brothers Karamazov* and the choice between freedom and security offered by adherence to religious authority." Guess how many lectures on Dostoevsky I had as an undergraduate.

In later chapters you will see how Watson nailed near-perfect narrative structure in writing what might be the most important research paper in the history of science, then eventually did the same with *The Double Helix*. It's not a coincidence—his level of success in science and the strength of his grounding in the humanities. He clearly has this powerful attribute of narrative intuition.

In graduate school I had two lecturers whom I idolized for their communications skills, Stephen Jay Gould and E. O. Wilson. Both legendary biologists. Both tremendous lecturers. Both thoroughly steeped in the humanities. I think part of my intrigue with them was just realizing what they possessed that I had failed to get in my training.

And let me tell you how long it took me to make up for this shortcoming. During orientation week at film school in 1994, three "older" graduate students (who were all probably ten years younger than me) spoke to us, giving us long-term advice about the program. They made it clear: when you graduate, the only thing Hollywood will value you for is your writing skill. They won't care if you've directed a great film—if movie companies want movie directors they generally look to music video and commercial directors who have intensively honed their visual skills. Film schools are the breeding ground for the more cerebral elements of filmmaking, namely, writing. So you'd better get to work writing three great screenplays that you can sell when you graduate because that is your only hope.

At that point in life, I was still a scientist, so I did what scientists generally do—I didn't listen. (Oh, whoa, no he didn't—did he just insult the entire profession of science? Actually, I defer on this note to the chief scientist of the Nature Conservancy and a member of the National Academy of Science, Peter Karieva, who, in reviewing *Don't Be Such a Scientist* for *Science*, said, "The failure of scientists as communicators is that they do not know how to listen, especially when it comes to the 'uneducated public.'")

I ignored this film school advice about the importance of learning to write, which basically means learning how to tell stories.

Telling stories requires that you have what I have labeled narrative intuition. All my classmates had already consumed mountains of novels as undergrads. They had the narrative templates imprinted on their brains. In contrast, I had consumed books about the phylogenetic tree. I could tell them more things about the difference between onychophorans and tardigrades than they could ever know . . . or, um, probably care to know. But I didn't have the same background in narrative.

So wouldn't you think that after three years of film school and five writing classes I would catch up with them? Nope. What about after writing and directing an entire comedy feature film after film school, writing screenplays that were represented by one of the top three talent agencies in Hollywood, and making short films about the oceans that were broadly popular (my public service announcement with comic actor Jack Black scored over $10 million in free air time). Still nope.

The entire process, from my start in making films to my final realization at a somewhat deep and intuitive level of the importance of narrative, would end up taking 16 years. It finally hit me in 2005 when I was making my documentary feature *Flock of Dodos*, a story I tell in *Don't Be Such a Scientist*.

That is how difficult and challenging this stuff is. Especially if you've already started molding your brain into the form of a scientist's. Is it hopeless? No. You just have to accept that it will take plenty of time and effort. The good news is, this narrative training is worth it because in today's increasingly communications-driven world, it's essential.

It's a Narrative World: Deal with It

There are many other basic rules about storytelling that can be counterintuitive and thus difficult to grasp. We'll get to more of them, but for now let me go back to the main point—that science

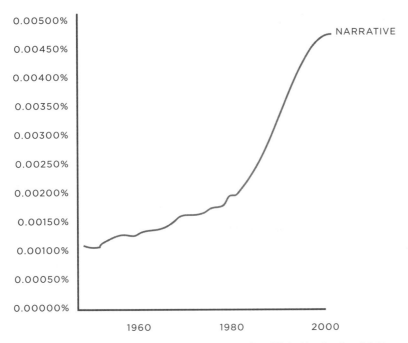

Figure 9. The frequency of the word *narrative* in recently published books. Google's N-gram Viewer allows you to search how often a term appears across most published books of the past few decades. This search result shows that, as information exploded in the 1980s, so did the use of the word *narrative*.

is stuck in a narrative world. As a result, scientists did not come up with their own unique mode of communication. Instead, they molded their language to that of the existing world.

Even though there has arisen an entire discipline called "science communication," I would argue that there is nothing unique to the way science is communicated. In fact, I fear that label actually sends the wrong message. I meet people who say, "I'm a science communicator," as if it were somehow unique. It's like saying you're a "baseball runner," or "soccer runner"—running is pretty much the same from one sport to the next. Same with communication from one discipline to the next.

So as I said, scientific journals began in 1665 with their own literary form, but eventually the Hegelian Triad took over and the structure of scientific papers adapted to it. Perhaps someday science will develop its own new template that does not conform to narrative and Hegelian dynamics, but for now there are no signs of that happening. Nor are there any signs of narrative playing a lesser role in our society. Just look at figure 9, which shows the frequency of the word *narrative* across recently published books.

Use of the word *narrative* erupted over the past couple decades, right in stride with the information explosion that took place in our society. Nowadays you hear it used constantly by news pundits, politicians, journalists, historians, economists—pretty much everyone who is trying to make sense of today's noise-filled world. On *The Daily Show*, Jon Stewart asked President Obama, "Do you buy into the Democratic party narrative?" But it wasn't always that way. I guarantee if you were to watch all the broadcasts of famed television news anchor Walter Cronkite, you would never hear him talking about the Vietnam narrative, or the space race narrative, or the China narrative. The word just wasn't used.

Today it is everywhere. Why is it so ubiquitous? I think the cause is information saturation. Narratives are stories that connect a series of events over time, creating large-scale patterns. When information becomes superabundant, it's only logical that people will look for higher-level patterns.

So if we accept that scientists are stuck in this narrative world and could use some help, then whom should they turn to for assistance?

2

AND the humanities
ought to help . . .

Here's a cool idea. You're a scientist on a college campus. You've realized you need help with narrative. You've also realized that over on the other side of campus are all these folks in the humanities who work with narrative all day long. Why not seek their help?

Before I tell you why not, let me first explain why, in a perfect world, seeking the help of humanities folks would be the perfect idea. Science actually has a lot in common with what "those people" do. Here's how.

Dude, It Really Is All the Same Story

Let's talk about "problem-solution" for both science and the humanities. It's pretty obvious that the scientific method is an exercise in problem solving. Here's a definition for the scientific method from a popular dictionary website:

> A method of investigation in which a problem is first identified and observations, experiments, or other relevant data are then used to construct and test hypotheses that purport to solve it.

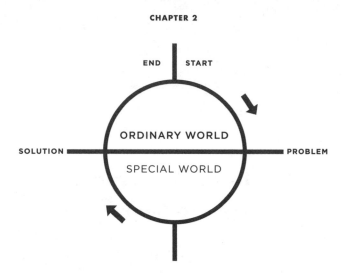

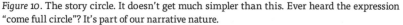

Figure 10. The story circle. It doesn't get much simpler than this. Ever heard the expression "come full circle"? It's part of our narrative nature.

Notice the words *problem* and *solve*. You address a problem in the form of a question, then you seek the answer. That's what scientists do all day, every day.

Now let's go back to story structure and Joseph Campbell—the guy with the mind of a scientist who crashed the story party in the 1940s. I mentioned that he identified a universal structure to stories, which he termed the *monomyth*. He also described the structure of a story as a circle. In Campbell's wonderfully simple conception, a story begins at a point in time (could be physical, could be mental) from which we head out, do some stuff, and eventually find our way back—essentially "coming full circle." This is the first point for grasping how a story works—one big circular journey. In the case of *The Wizard of Oz* (which I hold up as a model for broad storytelling), Dorothy leaves Kansas, goes to Oz, then eventually finds her way back home to Kansas—coming full circle.

The second starting point for grasping how a story works is to view a story as a journey between two worlds—the everyday world and the exceptional world. Campbell called these the Ordinary World and the Special World (as you see in figure 10).

The Ordinary World is where you, deep inside, yearn to live your entire life. It is your place of complete comfort, safety and security. You really never want to leave it. But things happen in life that throw you out of your Ordinary World (like getting yanked up by a tornado). Once you're taken out of your Ordinary World, you have only one major thing on your mind and one goal in life, which is to find your way back to that world of comfort. What's the very first thing Dorothy wants when she lands in Oz? To go home—back to her Ordinary World.

All of which means that as soon as you leave your Ordinary World, you automatically have a problem (how to get back), which means you need to undergo a journey in search of a solution. And there's your story—one big exercise in problem-solution. Same as science. Dude, same story.

When you begin to process this way of looking at a story, you also see that it's what you experience every day of your life. You start your day at home (your Ordinary World), you venture out to somewhere (the Special World), then eventually make your way back home—hopefully a little better for the journey. Then at night you go to the pub and tell your friends the "story" of your day.

So it seems obvious to head over to your university's English Department and shout out, "Can anybody here help us with our narrative problems?"

But I don't really recommend it. I'll tell you why.

3

BUT the humanities are
useless for this . . .

Okay, before I get lynched by a mob of professors in tweed jackets with elbow patches, let me hide behind my friend Jerry Graff. In addition to writing the hyperpopular textbook on argumentation *They Say, I Say* with his wife, Cathy Birkenstein, he has made a habit of tweaking the noses of academics. In particular, with his book *Clueless in Academe*, he pretty much slaps academia in the face with language far more harsh than I would ever use. His first chapter is titled "In the Dark All Eggheads Are Gray."

Wow. How much do I love Jerry Graff, my hero? I so thoroughly admire people who are well into their 70s yet still have as much fire burning in them as a young radical. It is possible to keep the mojo alive, and he's one of several septuagenarians I know who exemplify this. His book is an indictment of what everyone pretty much knows about academia. The place is a refuge for culturally detached blowhards, some of whom are good for teaching and research but often limited in their ability to function outside the ivory tower. Not all of them, but many.

The humanities in general have all sorts of unhealthy problems. For the last few decades there has been what is often labeled as the "humanities crisis." Benjamin Winterhalter addressed this problem nicely in a 2014 article in the *Atlantic* titled "The Morbid Fascination with the Death of the Humanities." In the article, Winterhalter examines the trend of bemoaning the humanities' sad decline. I first recall reading about this decline in the 1980s social crisis book *Cultural Literacy*, by E. D. Hirsch. It was an early cry of "What's this world coming to?" as college campuses in the new information era began to see the humanities displaced by the hegemony of science and technology. All across America students were replacing literature classes with computer science courses—trading one language for another.

Winterhalter tells of the predictability of *New York Times* op-ed pieces agonizing over the trend. The *New Republic* created the article tag "Humanities Deathwatch." In 2009 Mark Slouka unleashed a cry in the dark in *Harper's* titled "Dehumanized: When Math and Science Rule the School." (Note: I wrote to Slouka in 2009 about my first book. He rather grouchily assumed I was part of the science offensive and basically told me to get lost.)

In fact, the politicization of humanities programs is felt by many to be out of control. In *Clueless in Academe*, Graff talks about how the divide between traditionalists and progressives has widened to the point where even the individual programs are divided in philosophy.

In the late 1990s in one of the all-time great academic pranks, Alan Sokal, a physicist at New York University, submitted a phony paper to *Social Text*, an academic journal of postmodern cultural studies, just to show what a politicized mess the humanities are. Titled "Transgressing the Boundaries: Towards a Transformative Hermeneutics of Quantum Gravity," the article was intentionally a bunch of mumbo jumbo written to play to the politics of the editors. (In my head I can hear Butthead saying, "Heh, heh—he said hermeneutics.")

After *Social Text* published the article in 1996, Sokal revealed in another journal, *Lingua Franca*, that the article was a hoax. He re-ferred to it as "a pastiche of left-wing cant, fawning references, grandiose quotations, and outright nonsense." As you can imag-ine, the humanities folks didn't take kindly to Sokal's prank. It picked open the scab of the longstanding concern about "The Two Cultures," which was the title of a 1959 essay by literary figure/physicist C. P. Snow calling attention to the divide that had already arisen between the humanities and science.

Outrage over the Sokal Hoax sparked a 1997 symposium about the "culture wars," which was summarized in the 2001 book *The One Culture?* But by the admission of most of the participants, the debates largely involved academics blowing smoke past each other with little application to the real world.

It's sad really. Even great scientists don't seem to be able to effec-tively bridge the divide. During my first year of graduate school at Harvard, I was a teaching fellow for E. O. Wilson, one of the greatest biologists ever. He is widely known as "the father of entomology" as well as "the father of biodiversity" and has won two Pulitzer Prizes for his popular writing (though a lot of his popular writing is still on a fairly high cultural plane). He is a magnificent lecturer who, along with Steve Gould, made Harvard a dazzling place to study ecology and evolution in the late 1970s.

In 1998 he threw himself into this divide between the humani-ties and the sciences with his book *Consilience*—a word that means convergence of evidence. He called for the two cultures to converge. It was a great and mighty essay, which I read when it came out. But, the only reason I read it was money. At the time I was working for *National Geographic*'s feature film office in Hollywood, reading material for potential as movies. They paid me, and I plowed through it. Otherwise I never would have made it more than 20 pages.

As I say, it was a great and incredibly erudite essay about bring-ing the humanities and sciences together, but it was written in the language of intellectuals, for intellectuals. Which in the end is mostly what academics do—they talk to each other. That's fine for academia but is of limited practicality for society.

This brings us back to the humanities. I'm afraid they're a bit of a write-off for the sciences when it comes to addressing this serious, and I think urgent, problem of narrative deficiency. Scientists need help, but they must get it from people who go beyond theorizing and work in the real world. Which is what leads me to my overall recommendation . . .

4

THEREFORE Hollywood to the rescue

How's that for a shocker? The land that gives you zombies, vampires and Transformers also has what the science world needs. This is what I have learned in my 20-year excursion away from academia. Jerry Graff provides the core of my argument when he says in his book *Clueless in Academe*, "An old saying has it that academic disputes are especially vicious because so little is at stake in them."

That's the problem. There's never been that much at stake for academics once they have tenure. But in Hollywood, everything is at stake at almost every hour of the day. There's a common expression in Hollywood that is absolutely true and mentioned every day: "You're only as good as your last movie." Doesn't matter how esteemed you are—one stinker and you're cooked for a long time to come.

Let's look at this in scientific terms—specifically, evolution by means of natural selection. In nature, some environments provide harsh conditions, which are believed to cause natural selection to operate very quickly, while other environments are not as severe and therefore present weaker selective regimes. We can say that academics live in a "weak selective environment." Universities can be wonderfully inspiring and fun places to be, but they tend to suffer not just from detachment from the real world but also a willingness to nurture the weak and feeble.

I saw it as a professor. It was kind of beyond belief what some of the tenured professors in my department were allowed to get away with. One old guy was a hardcore alcoholic who literally—and I mean literally, like, this is the truth—canceled every other lecture in his animal biology course. He would show up and say, "It's a nice day. Everyone get outside and enjoy the weather." All the faculty knew he did this. There was nothing that could be done—he had tenure.

Which is no big deal. It's part of what goes on in academia, and there are great benefits that come at the cost of the occasional instances of tenure abuse. But overall, academia is simply detached from the real world and not the sort of incubator that is likely to help science with this very practical problem. It is a nurturing environment, but it is also the ivory tower and everyone knows what that means—this is nothing shocking I'm saying here. Hollywood is a very different story, as I have witnessed firsthand over the past two decades.

Hollywood is a brutal, rotten, vicious, heartless place. One of my favorite memories of the place is of a friend from acting class who did a movie with comedian Rodney Dangerfield at the very start of her acting career. She was in her late twenties, a funny, charming actress, and I visited her on the set numerous times. Rodney instantly bonded with her. Between takes he would come over in his bathrobe and sandals (which he wore whenever he wasn't in a scene) with a cigar and cocktail and say to her, "Get out now, kid. It's a ROTTEN business." And he wasn't joking—he meant it. She and I quoted that line over the years as we both encountered endless scummy behavior in Hollywood.

What? You want a more specific example? Are you going to cite the rule I already told you about, that "the power of storytelling rests in the specifics"? Okay, here you go—brace yourself. This same actress a few years later came down to the wire for the lead role in a Showtime series. She got to the final audition, felt she nailed it, and everyone present told her she was amazing. But then they gave the part to the other finalist—Rebecca Gayheart. My friend was stunned, as was her manager, who called the producers then reported back, "They went with Rebecca Gayheart because she has greater name recognition." Guess where the name recognition came from . . . literally . . . in the year before, in 2001, she struck and killed a nine-year-old child with her car in a crosswalk. She was all over the news for it. She had name recognition from a terrible thing, but the producers didn't care. That's show biz.

They say that in Hollywood your competitors don't hope you fail, they hope you die. Which is true. I remember an actor friend inviting me to the funeral of a casting director. He said he only met her once but funerals are great networking opportunities in Hollywood. The place is heartless.

I can't believe the levels of deadly sin I've seen in Hollywood—jealousy, envy, greed, lust, gluttony—it's a mess. There are countless books written about it. My all-time favorite is the surprisingly powerful, yet wistful, *You'll Never Eat Lunch in This Town Again* by Julia Phillips, the Oscar-winning producer of *Close Encounters of the Third Kind*.

But of course you already knew, or at least suspected, as much. When I was still a professor, the *Chronicle of Higher Education* wrote an article about my short films on sea creatures. When I got ready to leave for my new career, I contacted the writer. She wrote a short follow-up piece sarcastically titled "Professor Leaves Academia for More Nurturing Environment . . . Hollywood." Even they know how rotten the place is.

As a result, Hollywood, unlike academia, has been the sort of "intense selective regime" that produces rapid evolution. Every weekend the box office totals of all the movies are published. It might as well be a professional death list. If you made a big movie, it opened, and it's not making back its money, you're dead. If Charles Darwin were around to read the *Hollywood Reporter,* he'd be doing back flips every Monday morning shouting, "Survival of the fittest!"

Over the years I've spent hours on the phone talking with film school classmates and filmmaker friends after their movies have collapsed, feeling their pain as they talk about the deaths of their moviemaking dreams. And almost always, at the core of the disaster, was weak narrative dynamics in the form of a failure to tell a good story. The selective pressure is relentless, and there's almost no collective memory. You make a flop, you go to "movie jail," where they won't let you make any more movies for somewhere between a while and forever.

There is no security, and seniority is a liability, not an asset. It's like a herd of hooved mammals on the Serengeti surrounded by lions, cheetahs and humans with high powered rifles. Age—rather than a badge that earns you authority and respect as it does at least somewhat in academia—in Hollywood is something to be hidden at all costs. Not just a matter of being a feeble animal; more like being in the wrong political party during a military coup. As if there were Gestapo who come around and check your papers for your age, then haul you out of Hollywood if you're over 40. (By the way, keep in mind I *started* film school at age 38.) This is the driving force for most of the facelifts, hair transplants, Botox injections and antiaging procedures there. You'd do the same if your survival depended on it.

Academia is a luxury resort by comparison. If you have a good run early on and hang in there, you are rewarded with tenure, which makes you untouchable for life. Try explaining the concept of tenure to a group of ravenous and emaciated Hollywood screenwriters. They'll probably attack you with spears and arrows.

This is how it's been in Hollywood for a century, yet lots of smart people have figured the system out well enough to survive. They know that to survive as a moviemaker you have to tell good stories, to tell good stories you have to understand narrative at a deep and intuitive level, and to do that you have to have studied, analyzed and refined narrative dynamics to a science. Which they have.

From George Lucas using Joseph Campbell to structure *Star Wars* to Christopher Vogler breaking down what Lucas did for the industry in *The Writer's Journey* to Blake Snyder putting it into the broadest and most vacuous form with *Save the Cat*, they have figured it out. They have refined the practical, real-world application of narrative far better than anyone else.

What it all comes down to is that Hollywood has at work this eternal selective agent called money. If you don't figure out how to put narrative dynamics to work for you in the making of money, you die. That is indeed natural selection, in all its ruthless splendor.

And this is why I now make my pitch, my plea, my proselytization that it is time for the world of science to set aside the cultural divide. I'm asking you to turn a blind eye, hold your nose or do whatever it takes simply to use these Hollywood lunatics for what they have to offer.

Again, it's not about their big budget, brainless, tent-pole movies that will forever prioritize storytelling over the truth. Nor is it about trying to dress, talk or act like them. It is about the knowledge of narrative that underpins it all. That knowledge is cool. And powerful. And science needs it. So here we go now, to consider the idea of a different world—one in which science is as savvy with narrative as Hollywood.

III

ANTITHESIS

Skype can be great. You feel as if the other person is in the room with you. It lets you communicate with subtlety. So it was the natural means of communication for connecting with the "aren't we special" scientist after telling Megan I was back on for the sea level rise panel.

This scientist and I have known each other for more than a decade—the whole bit of tension really was kind of silly. It took less than a minute for us to get past our emails and set to work with the narrative tools I have developed for the Connection Storymaker workshop.

I began by asking him if there might be one word at the core of the entire issue of sea level rise—one word that captures the essence of the issue. He thought for a while, then finally gave up, saying, "I'm sorry but I can't give you a single word for the entire issue, but . . . can I give you three words?"

I said that's close enough, let's go for it. The words turned out to be great, and we shaped them into the new title for the panel. We replaced the original, rather dull title of "Responding to Sea Level Rise" with "Sea Level Rise: New, Certain and Everywhere," which was much more specific and thus more powerful.

Then we went to work on each word with the ABT Template. Actually, we went to work on a lot of stuff. Then—cut to four months later—we found ourselves in a ballroom at the end of our presentation enjoying loud and prolonged applause. Our panel was a raging success (as I will describe in detail later in this section) with the 1,000 eager minds Megan had promised. And a month later *Science* published my letter about how the ABT had transformed the event. Bottom line: the tools work.

So it's time to talk about "what could be" in the world of science. Again, I'm not suggesting that these tools are a magic bullet or a panacea (calm down, negators), but they can solve the narrative deficiency problem.

To achieve this, the narrative tools need to become a fundamental part of science. Not an add-on bonus for postdoctoral scientists, or graduate students, or even upper-level undergraduates, which seem to be the groups I typically get called in for right now. Narrative needs to be taught from the start of science at the undergraduate level.

This section of the book is the core of our journey. It is the middle part of the story. It is where things happen. If we want to look at it in terms of Aristotle's structure, there are basically three episodes for us to go through in this section. They are the three elements of the WSP Model—the word, the sentence and the paragraph. For this section I'm going to use the MR labels of the IMRAD.

Notice that I'm not using the ABT structure for this section. That structure works well for the Introduction and Discussion (Thesis and Synthesis), which are more subjective and along the lines of argumentation. This middle section is just nuts and bolts, the more objective material about the journey—just the straightforward reporting of what happened. Hopefully I've pulled you in with the initial ABT structure of the Thesis and gotten you sufficiently engaged such that you're interested in following the "events" described here. Then in part 4, Synthesis, I'll return to making my argument and telling my story using the ABT.

5

Methods: Narrative Tools

The WSP Model

It's Time to Talk Substance in Our Age of Style

The subtitle of *Don't Be Such a Scientist* was *Talking Substance in an Age of Style*. In that book I focused much more on problems than on solutions. Which was fine. Notice the subtitle wasn't *How to Talk Substance*. My purpose in writing the book was simply to shine a light on the challenge of talking substance in a world so overrun by style. I didn't know enough specifics on narrative then to offer up suggestions for a remedy.

Over the next four years Dorie Barton, Brian Palermo and I created our Connection Storymaker Workshop, the essence of which we pulled together into our book, *Connection: Hollywood Storytelling Meets Critical Thinking*. The workshop focused on the power of narrative and over time resulted in a set of narrative tools. In retrospect, I see that the workshop constituted my journey in search of the answer for how to communicate substance.

This book is not about style. Things like the use of humor, emotion, plain language, clever metaphors and snappy dialogue are all elements of style, which are also essential parts of effective communication. But for communication to connect at a deep level, to unify large audiences and have a lasting impact, it has to begin with substance. Basically, shape the information first, then add the style elements. Narrative is the substance of what you have to say.

Time for Lift Off

Referring back to the "see it, shape it, say it" process that Medawar identified, it's now time to begin the shaping process. This is what narrative requires. As I have said, scientists detest the idea of shaping, but I will forever meet their objections with a single acronym—IMRAD. That template forces scientists to shape their information. If narrative shaping was good enough for scientists a century ago, it's good enough for today's scientists.

My approach to narrative shaping is the WSP Model. I first presented it in *Connection*. Here I apply it more specifically to the world of science. It's about shrinking the narrative core of your story down to one word, one sentence and one paragraph as a means of developing and strengthening its structure.

There is a tool for each process. The tools are called templates—sentences with blanks to be filled in. For example, this sentence is a template you could use to start a conversation with your friend: Hey, _____, I need to talk to you about _____. Just fill in the blanks.

The key to storytelling is finding the narrative core of what you or someone else is trying to say. Once you've shrunk the story down to its smallest bits and found the core structure—then you can set to work expanding it back out.

I'm sure you know all the old jokes about short versus long communication. Like the saying about a letter: "I would have written less but I didn't have the time." Or about a talk: "If you want me to speak for an hour I'm ready now; if you want only ten minutes I'll need a week to prepare." Brevity, though the soul of wit, takes time and energy. But the process can be helped along with the WSP model. Each part of the model works differently in terms of immediate (proximate) versus long-term (ultimate) effectiveness. Let's take a look at this basic dynamic before we start.

Proximate versus Ultimate Value

The three tools of the WSP model have different proximate versus ultimate value, as you can see in figure 11. The Word Template and the Sentence Template have great proximate strength. You can pick them up, put them to work and within minutes have a clearer grasp of the story you want to tell. You can master them quickly.

The Paragraph Template is different. It's the big kids' tool. It's for the long haul. Most people in Hollywood wish they had it mastered, but they don't even have a beginning knowledge of it. The learning curve for facility with the Paragraph Template is much longer than for the other two. You may think you get it immediately when you fill in the blanks, but the fact is you will probably need a very long time to achieve meaningful results. Ultimately, though, the Paragraph Template will take you much further in developing your narrative skills than either of the other two.

Also, for the Sentence Template (the ABT), the short-term versus long-term benefits are very different. For the short term, the

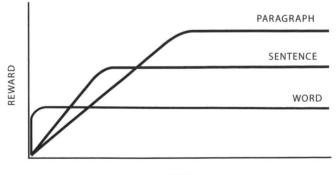

Figure 11. Return over time for the three WSP tools. The Paragraph Template is just a silly toy at first, but with time it can make you into the sort of storyteller Joseph Campbell would have admired.

ABT can help you find immediate narrative structure in a "pile of sundry facts" (a phrase we'll come across shortly). But in the long run it gives you the ultimate prize—narrative intuition. If you work with it enough and make it second nature, you will begin to develop an intuitive sense for what is wrong with poorly structured material and how to fix it. It's that intuition that everyone needs . . . ultimately.

These days I focus primarily on the ABT Template in my workshops. It's the Goldilocks thing. The Word Template is too short and of limited range; the Paragraph Template is too complicated to understand quickly and takes a long time to master. But the Sentence Template is just right—quick to learn, immediate in its value. For this reason we will spend a lot of time with it here.

The goal, as I mentioned at the outset, is intuition with regard to narrative dynamics, which I am calling narrative intuition. There's been a lot written about the power of intuition in recent years. My favorite work is Malcolm Gladwell's *Blink: The Power of Thinking without Thinking*. He opens the book with the observation that a good art forgery detective can spot a fake in an instant, yet will need a lot more time to explain exactly what criteria led to that conclusion.

Intuition is defined by at least one dictionary as "a feeling that guides a person to act a certain way without fully understanding why." It is the art side of communication. Most of us were not born with much intuition—we achieve it through experience. In his book *Outliers: The Story of Success*, Gladwell offers up the somewhat arbitrary number of 10,000 hours as the amount of experience needed to move a complex skill from the cerebral (memorized) to the visceral (intuitive).

This idea is also the gist of a follow-up article Gladwell wrote in 2013 for the *New Yorker*, titled "Complexity and the Ten-Thousand-Hour Rule." Talking about research on his proposed 10,000-hour rule, he said, "The ten-thousand-hour research reminds us that 'the closer psychologists look at the careers of the gifted, the smaller the role innate talent seems to play and the bigger the role preparation seems to play.'"

I am in agreement with this, and you should take it to heart if you feel like you're not that good with narrative structure. Just get to work. You may not be able to find 10,000 hours for it, but even the 10 hours I will eventually propose in my Story Circles concept in this book's final chapter will move the needle for you.

The solution to the narrative deficiency problems of science rests in what Gladwell terms "preparation"—meaning a great deal of repeated practice, plain and simple. No easy fix, just a lot of hours spent doing things the right way—learning fundamentals, just like an athlete.

There is much I can teach you quickly, but intuition takes time and experience. As Alan Alda, the beloved comic actor who found a second career as a trailblazer in helping scientists communicate, says in his book *Things I Overheard While Talking to Myself*, "Good communication can be taught. But for it to have some lasting effect—for it to become a part of someone's core—I think it has to be taught systematically, and over time." All scientists have some feeling for narrative, but they need the deeper level of intuition.

Tools for the Masses

Now it's time to get back to the "Dude, it's all the same story" message. I am writing this book for the world of science because that is where I hold my first allegiance, but make no mistake, these tools are for everyone.

Four thousand years of human diversification really isn't that much. It was only about 70 years ago when Joseph Campbell pointed out how, around the world, we're all still telling the same basic monomythic story. Nothing changed in those 4,000 years. And nothing significant has changed in the past 70 with regard to storytelling (a point I will make later in reference to what still makes for successful Super Bowl commercials).

I work with a wide range of clients on storytelling. In the past year I have run my workshop for accountants (Deloitte), safety workers (National Safety Council), business professionals (Society of Marketing Professionals), and lots of science and biomedical groups (National Institutes of Health, Centers for Disease Control and Prevention, American Geophysical Union, American Association for the Advancement of Science, Society of Hospital Medicine). In the beginning I used to be a little daunted entering those different realms. I know virtually nothing about accounting, business or law. How could I offer anything useful to them?

But what I experience with these various groups ends up being the true Joseph Campbell thing. I get to see that it is indeed all the same story. Each group's content is different, but I'm there for the structure and delivery, and that is the same everywhere. I'm like a construction worker—doesn't matter if we're building a bank, a hospital or a courthouse—I'm just there for the structure part of it.

Aren't Templates for Toddlers?

Is it possible to be "too simple" when it comes to communication? Of course, but when it comes to a profession like science, which constantly suffers from being too complex for the public, being too simple (all else equal, meaning the information is kept accurate) is a small worry.

Nevertheless, because templates seem so elementary, many people get a feeling of "that's for children" as soon as you begin to talk about them. I got a taste of this at a meeting where I gave the opening plenary. At the reception following my talk a friend pulled me aside and said, "What I'm hearing is the scientists think your ABT thing is neat, but . . . they also feel it's too simple for their communications needs."

I'm not quite sure what to do with that comment. Simplicity is the essence of effective communication. If you don't grasp the importance of it, you don't grasp effective communication.

Unfortunately, because of the social dynamics of the science world, scientists are often allowed to present their work in ways that are incredibly complex and confusing, and no one complains. But it shouldn't, and doesn't, have to be this way.

This is where templates come in. You remember them from elementary school and games like Mad Libs. But just because children use such devices, does that mean templates are beneath adults?

Jerry Graff and Cathy Birkenstein take on exactly this question in *They Say, I Say* with a section titled "Okay, but templates?" Their book is about argumentation, and they refer to the key elements of making an argument as the "moves" that are needed, which their templates provide. They hit the nail on the head, saying, "While seasoned writers pick up these moves unconsciously through their reading, many students do not." And this is why templates help.

The criticism they're addressing is exactly what I encounter—seasoned veterans dismissing the ABT as trivial and for children. That's fine for them, but first off, most people aren't as adept at narrative as veterans, and second, I guarantee you even a lot of the veterans could benefit from practicing with these templates.

Professor Cartman

Guess who probably is a superstar with templates—Trey Parker, co-creator of *South Park*, who was my initial source for the ABT Template. One of my all-time favorite episodes of the show is "Funnybot," in which a group of Germans create a robot that solves the problem of comedy mathematically by coming up with templates for jokes. Like this one from the robot's stand-up routine for the school kids: "Don't you just hate doing [activity]? Me, too, man, I hate [activity]. Honestly, I hate having to do [activity] more than I hate having to do [name of a person] in his [bodily orifice]. Awkward!" For this one the activity blank was filled in with "homework," the name of the person was Bryant Gumbel, and I'll spare you the bodily orifice.

It is indeed awkward, as Funnybot says, but it also works. Why wouldn't it? Joke telling is the same as storytelling. Setup/twist/punchline is the same as thesis/antithesis/synthesis. All the same story, dude—duder—el Duderino.

Bottom line, if you think a template is "too simple" for your communications needs, you're probably part of the problem. And if you think you're addressing the same issue as those who complain about television sitcoms being "too formulaic," you're not looking at the communications problems properly. Here's what I mean.

Form not Formula

The problem of material being overly formulaic arises when the material itself is devoid of content. For example, if the characters in a sitcom are shallow and vacuous because of a shortage of interesting details about them, then when an old boyfriend shows up asking to borrow money in this week's episode, you're immediately going to think of all the episodes of everything from *Modern Family* to *Friends* to *The Dick Van Dyke Show* that involved former lovers showing up, asking to borrow money (I'm not sure such episodes exist for those shows, but you get the idea). But if enough information has been presented to establish the characters as deep and interesting in their own right, then you will be drawn into the story without ever thinking of the similarities to previous shows.

Christopher Vogler's iconic book *The Writer's Journey: Mythic Structure for Writers* addresses the form versus formula issue directly. First published in 1998, it's in its third edition. But the book is a lightning rod for all who feel that Hollywood produces "formulaic garbage" that makes the world a lousier place. He takes on such critics at the start of his book:

First, I must address a significant objection about the whole idea of The Writer's Journey—the suspicion of artists and critics that it is formulaic, leading to stale repetition. Some professional writers don't like the idea of analyzing the creative process at all, and urge students to ignore all books and teachers and "Just do it." Some artists make the choice to avoid systematic thinking, rejecting all principles, ideals, schools of thought, theories, patterns, and designs. For them, art is an entirely intuitive process that can never be mastered by rules of thumb and should not be reduced to formula. And they aren't wrong. At the core of every artist is a sacred place where all the rules are set aside or deliberately forgotten, and nothing matters but the instinctive choices of the heart and soul of the artist.

But even that is a principle, and those who say they reject principles and theories can't avoid subscribing to a few of them: Avoid formula, distrust order and pattern, resist logic and tradition.

Artists who operate on the principle of rejecting all form are themselves dependent on form.

So is there a concern about scientific communication becoming too formulaic? The IMRAD template already makes it formulaic, but you don't hear anyone complaining about that. The problem of nonstructure and excessive complexity is so severe in the communication of science that it would take a cataclysm of templating even to begin to override it. So I will mostly defer to Vogler, Graff and Birkenstein. It doesn't worry them; it doesn't worry me.

Also, scientific communication has a long, long way to go before "being too similar" becomes a serious concern. You'll see this when I take to analyzing abstracts with the ABT Template. There's currently no sign of excess uniformity being a problem. Not by a long way.

Okay, time for our first template.

6

Methods: Word

The Dobzhansky Template

This first template, the Word Template, is for finding the central theme of your material. It's not just about summarizing everything in a single word (or singular phrase)—it's deeper than that. It's about thinking long and hard to find the one word that is at the core of it all. I have dubbed my Word Template the "Dobzhansky Template" because it is an adaptation of a famous quote from Theodosius Dobzhansky, a geneticist who I doubt even knew he was putting his finger on the whole idea of narrative.

Dobzhansky was one of the most important geneticists of all time. He immigrated to the United States from Russia in the 1920s. His 1937 book *Genetics and the Origin of Species* was a central component of Modern Synthesis (the merging of genetics with natural selection). But he was much more than just a researcher.

Dobzhansky fostered many outstanding graduate students who would go on to be leading lights of genetics for a generation—especially in population genetics. One of them is Richard Lewontin, who was one of the most accomplished evolutionists at Harvard when I was there and was also Stephen Jay Gould's longtime colleague in their efforts to challenge genetic determinism. Francisco Ayala, a recipient of the National Medal of Science, is another.

Many of Dobzhansky's students are still around, and I spoke to some about him, including Wyatt Anderson, who was his first student at Rockefeller University in 1962. He and Dobzhansky spent many years traveling in the western United States collecting the fruit flies they both studied. "He was truly charismatic, very thoughtful, and interested in biological philosophy," Wyatt told me. "He had a warm personality. He liked music, art, and was an expert horseman."

I wanted to know about these traits because I had a hypothesis— that Dobzhansky must have had an exceptional understanding of human nature. It seemed to me this had to be the case. It hit me when I looked at the famous line he is known for, which is both simple and complex. Dobzhansky first mentioned the line in an essay published in *American Zoologist* in 1964. It appeared in many of my introductory biology and evolution textbooks when I was a student. Here's what he said:

Nothing in biology makes sense except in the light of evolution.

At first glance it seems simple—just an observation about the importance of knowing about evolution. But there's a secondary, and I think more important, dynamic to what he said. The statement also turns out to be a pathway to a deeper and more functional understanding of what is meant by narrative and how it works.

It's kind of funny and fitting that this statement comes not from some obscure microfield of science but from evolution—the greatest and most overarching subject in biology. Before there were birds and bees, there was evolution, all the way from the first sparks of life on the planet.

Dobzhansky is saying that evolution is itself the "story of life." A story is about change (about a journey). Evolution is the mechanism of change that produces the patterns of change. It is the narrative of life. You can look at all life on Earth, but for it to really make sense, you need to know this story of change—this narrative—which is evolution.

The first thing to note is what one of my former evolutionary biology colleagues is quick to point out (ah, the critical minds of scientists). My friend doesn't like the quote because it's just not right. He points out that there are plenty of things in biology that can make sense to you even if you are completely oblivious to evolution. There are molecular biologists producing DNA sequences all day long, which make total sense despite the biologists' lack of understanding of evolution. As a result, he objects to the word *nothing* in the quote.

He's right. Somewhat. Actually, it only shows us how human Dobzhansky was. Humans are driven to tell big stories. Had Dobzhansky said, "*Hardly anything* in biology makes sense except in the light of evolution," he wouldn't have had the impact he did by saying the more extreme word, *nothing*. Like all humans, he wanted to tell a big story, even if it wasn't correct 100 percent of the time. He probably thought it was close enough.

He also wanted to be compelling and concise. The word *nothing* is a superlative (a word of extreme), making it both more compelling and more concise than the phrase "hardly anything." It's similar to how the moviemakers revised the astronaut's line to be "Houston, we *have* a problem."

But even so, I'm taking Dobzhansky's quote in the direction of "finding the narrative," which is basically seeing the forest for the trees. You need to be able to stand back and look for overriding patterns and not get caught up with the noise.

The Dobzhansky Template: Finding the Narrative

So this single, simple sentence derived from Dobzhansky's quote becomes a template to use in your initial efforts to "find the narrative" of any given topic. Dobzhansky never saw the quote as a narrative tool, but I definitely do. Here it is as a template:

Nothing in _____ makes sense except in the light of _____.

Start applying it to other topics and see if it works. I mentioned it to a geologist. He immediately finished the sentence with "plate tectonics." Here was his thinking: you can look at earthquakes, volcanic eruptions, subduction zones—you can even find fossil seashells on the tops of mountains—all of which will be fascinating but never really make sense except in the light of . . . plate tectonics. That is your narrative for geology—the one factor that explains virtually everything. "Nothing in geology makes sense except in the light of plate tectonics."

This is sort of the definition of a narrative. It unifies a whole bunch of seemingly disparate pieces of information. Suddenly, armed with the knowledge of plate tectonics, the earthquakes, the volcanos, even the seashells on the mountaintop make sense.

Here's another example. A friend of mine in his youth had all sorts of joint pains, headaches, gastrointestinal problems—nonstop. The doctors could never make sense of it and never gave him any sort of unifying diagnosis. They only came up with treatments for each symptom. But when he was 33 he was finally diagnosed with a genetic disease, Ehlers-Danlos syndrome, in which a set of defective genes cause poor collagen production, resulting in weak connective tissue and creating the entire suite of my friend's symptoms. In an instant—in a single moment—he was able to fill in the Dobzhansky Template, "Nothing in *his life* made sense except in the light of *this genetic disease*."

It's the same thing with King George III of Great Britain, who suffered a range of physical ailments, eventually culminating in complete madness. Historians think it was the blood disease porphyria—the single element that gave rise to the entire narrative of his life (encapsulated, for example, by the film *The Madness of King George*). Nothing in his life made sense except in the light of that disease.

Now try applying it to your life or your research program or your lousy tennis game. ("Nothing in my tennis game makes sense except in the light of my damaged ankle that throws off everything.") What's the one factor that explains and embraces all? That's the narrative.

Nothing at *Apple Computers* makes sense except in the light of *innovation*. That's pretty much their narrative, and guess what, it's also their "brand," because narrative and brand . . . yep, you guessed it, dude. All the same story.

An environmental lawyer friend came up with one for her work on climate change: Nothing in *California climate change* makes sense except in the light of *loss*. The Dobzhansky Template produced the one-word theme of "loss" that she realized is at the core of every talk she was giving about the state of climate change in California. All of her talks were about drought and wildfire resulting from climate change. But more broad and unifying was the fact that her talks were all about loss—what the state is losing and will continue to lose because of climate change.

You can see how empowering this template can be if you're able to make it work. In my friend's case, she can keep coming back to her key word in a talk, saying, "What we're talking about here with all of these examples is loss—what we will be losing in the near future because of the changing climate."

Try completing this template with your favorite high-quality movie. If it's a movie that has depth and complexity, and reaches a large audience, you should be able to fill in the blanks. For example, one of my all-time favorite dramatic movies is *Ordinary People*, which won four Oscars, including Best Picture. It's a powerful story and falls right into the template with "Nothing in the story of that family makes sense except in the light of the death of their son." See what I mean with that? It is the one element that explains everything wrong with the family. Their failure to deal properly with the son's death became the source of endless problems.

Conversely, a lot of lousy, shallow movies suffer from the absence of this deeper, unifying element. When people walk out of a movie and say, "I don't even really know what that story was about," they are basically saying they couldn't fill out the Dobzhansky Template for it.

If you can fill in the blanks of this template (and by the way, it's not the case that you always can), suddenly you're greatly empowered. This becomes your "message," enabling you to engage in messaging more effectively. That's what "messaging" and "staying on message" is all about—hitting the theme from multiple angles.

It's also a technique that many actors use in "breaking down" a script. They read a scene then ask themselves, what is the one word that is at the core of the scene? Is it trust, love, deceit, betrayal, loyalty, endurance?

I was introduced to the underlying concept of this approach while promoting my movie *Flock of Dodos*. After doing a mediocre and rambling interview on NPR's *Talk of the Nation* about the movie, I had a chat with my very savvy sales representative, Jeff Dowd (who, by the way, was the guy *The Big Lebowski* was based on, for reals). He had mass communications experience from working on national political campaigns such as John Kerry's run for president in 2004.

He ran me through a simple exercise. He asked, "What is the one word that is at the core of your entire movie?" I replied, "Evolution." He said no. I said, "Creationism." He said no. I said, "Controversy." He said no. I said, "Okay, I give up, what's the word?" He said, "The word is *truth*." At the core of your movie is the struggle over what is the truth, who is in control of the truth, how can we make sure the truth prevails." He was right. Those other words were relatively shallow and inert. *Truth* has a human dimension to it, making it very powerful. It is, in fact, the "narrative" of my movie.

He then went on to say, "So now you know your narrative and you can use it for messaging. In the future, whenever you come up short in an interview, you can always fall back on this by saying something like, 'At the core, what this movie is about is the truth—who is control of it, how can we make sure it prevails . . .' "

It worked. From then on I never had another rambling, contorted, directionless interview. I had a message. This is what's meant when people talk about "staying on message." It's all about knowing the narrative. But there's more.

The Important Stuff: Dobzhansky's Second Part

Now that you've got the basic template, it's time to dig deeper. There's more to what Dobzhansky had to say. In a 1973 paper he split his idea into two parts. The first part is pretty much the same as the short quote:

> Seen in the light of evolution, biology is, perhaps, intellectually the most satisfying and inspiring science.

But then he goes on to add a second part:

> Without that light, it becomes a pile of sundry facts—some of them interesting or curious but making no meaningful picture as a whole.

This is where the real communications gold lies. Let's take a close look at what Dobzhansky is saying with these two pieces. In the first part he identifies evolution as the "narrative" of biology. In the second, he describes what happens if you don't have a narrative.

Notice that Dobzhansky doesn't say that all is lost if you don't have "that light" (meaning the narrative). You still have a bunch of information (a pile of sundry facts). Furthermore, some of the information may be "interesting or curious." The only problem is that ultimately, without the narrative, what you have makes "no meaningful picture" as a whole.

This is pretty much your whole narrative dynamic, all in one short quote. The second part of the quote describes the vast majority of the gee-whiz programs about science you see on television. Most are packed with tons of pieces of exciting information, many of which are without a doubt interesting or even curious. But ultimately, there just isn't a deeper narrative present, and thus no meaningful picture as a whole. Even some of the very biggest, most exciting-est science TV series ever lack this deeper narrative.

Moreover, this is the very trap that so many scientists fall into with when delivering research talks. They present a whole bunch of pieces of information that are definitely interesting, and some of which are downright curious, but in the end they are just a pile of sundry facts. The talk fails to make clear where the facts fit into the bigger picture and how they are helping to "advance the narrative."

This was okay in a world short on information. Up until the 1970s it was no problem. I began college in the 1970s. Nobody talked about there being too much information. Universities were seen as oases of knowledge in a desert lacking in information. They were "beacons of light" where you went to find treasure troves of information. But that all changed in the 1980s when the information tide suddenly reversed.

Today we are a culture awash in information. Most people have a little voice playing constantly in the back of their minds saying, "Why do I need to know this?" whenever they are being told something. The Dobzhansky Template is a tool that helps you answer that question. Kind of like this . . .

"I'm going to tell you about how evolution works, and you need to know this because nothing in biology makes sense except in the light of evolution."

"I'm going to tell you about this disease because nothing in your life makes sense except in the light of this disease."

This is the pathway to truly powerful communication. Find the narrative and you've found the key to everything.

Your Narrative Theme

What we're talking about in general terms here is called "theme" in the world of literature and creative writing (yep, all the same story). Remember the point I highlighted from Terkel's book *The Good War*? That most people who lived through World War II realized it was the most meaningful period in their lives? For those people you could probably fill in the Dobzhansky Template this way: "Nothing in their lives makes sense except in the light of what they experienced in World War II." It became the theme of their lives.

Most good writing teachers will tell you that meaningful writing starts with having a theme. They will ask, "What are you trying to say here?" This is the question I find myself asking workshop participants over and over again. They read a brief summary of a story they are working on, then I start in with that question. I explain that if the other workshop leaders and I know what you're wanting to say, we can help you say it better. But if you don't know yourself, it's kind of hard for us to help you. Often the participants have a story that is really cool, and interesting, and fun, but . . . they don't really know what it says or how it would be useful in any practical sense. Which is fine. It's just not as meaningful as it could be.

There might not be a simple answer that completes the Dobzhansky Template for your story, but you'll never know until you give it some thought. Also, keep in mind that the more "human" the term you come up with, the more dramatic, and thus more powerful, the message possible. In the case of my evolution movie, the words *evolution*, *creationism* and *controversy* were all largely informational terms. But *truth* is a core human value. You're hitting on Superman material with that ("Truth, Justice and the American Way!").

Just Because It Happened to You Doesn't Mean It's Interesting

Let's go back to the fundamental question of "Why do I need to know this?" It's kind of an ugly question, but having an answer for it makes for better communication. My Connection Workshop co-instructor Dorie Barton is fond of saying, as she reads bad movie scripts, "Just because it happened to you doesn't mean it's interesting." It's a horrible thing to say, but it's actually worth storing in the back of your mind.

You may have emerged unscathed from a five-car pileup accident where all four of the other drivers were members of rock bands. That's kind of fun, and we might enjoy hearing a couple sentences about it, but after a while, if you are going on and on with all the details, "Why do I need to know this?" will become the pertinent question. Someone should say to you, "Okay, that's really unlikely, and it's cool that it actually happened to you, but aside from being mildly amusing, why is it interesting to me? What does it say about larger issues? How does it help create a meaningful picture overall?"

Basically, in the end, we really don't care that it happened to you. We need it to have some deeper significance and meaning if you're going to eat up a lot of our time with it. What is the larger narrative?

Why Should We Care about Your Grant Proposal?

This question points to the real relevance of the Dobzhansky Template to your science career—writing grant proposals. It's the dreaded response I always hated hearing from my program officers at the National Science Foundation as I sought feedback for my rejected proposals. They would say, "Why should we care about metamorphosis of holothurian larvae?"

I wanted to reach through the phone and slap them. And I would usually, stubbornly, foolishly give them some speech that was really only about the idea of knowledge for knowledge's sake ("The world needs to know this!"), which just doesn't answer the question. I was terrible at writing grant proposals. One of the great reliefs I felt in leaving my science career was the idea that I would never, ever again have to hear that dreaded question, "Why should we care?" And yet . . .

It wasn't but a year or so later I found myself pitching movie ideas to producers in Hollywood, and guess what they would say . . . "Why should we care about the story of a group of marine biologists studying a coral reef?" Argh. Honest to goodness. I wanted to slap them, too. But I think it was at that point that I finally began to actually hear this question and start to grasp what it means.

Granting agencies want to fund important work. The definition of important research is that it has the potential to "advance the narrative" for a given subject. Filling in the blanks of the Dobzhansky Template gives you the materials to put your work into context and make the case to justify funding. If you can tell the granting agency, "Nothing in mitochondrial genetics makes sense except in the light of my current work on _____," you're probably going to get their attention.

Big Data and the Brickyard

Knowing your narrative is more important today than ever, given the ocean of information in which we are now awash. However, scientists have actually been thinking about this problem for decades.

In 1963, *Science* published a simple, very nonliteral and almost landmark short essay titled "Chaos in the Brickyard" by a medical researcher at the Mayo Clinic, Bernard K. Forscher. It is an allegory for the state of the science world. Only one page, but fascinating to read, in part because I doubt *Science* would publish such a completely nonliteral piece of writing today.

It begins in the classic style with "Once upon a time, among the activities and occupations of man there was an activity called scientific research and the performers of this activity were called scientists. In reality, however, these men were builders who constructed edifices, called explanations or laws, by assembling bricks, called facts."

You can break the story down into its simple structure using the ABT Template that we're about to get into. It is about how bricks were made AND buildings were built, BUT then the builders became obsessed with the making of bricks, regardless of need, producing a superabundance of bricks. The essay goes on to say, "And so it happened that the land became flooded with bricks. [THEREFORE] It became necessary to organize more and more storage places, called journals . . ."

You see what the story leads to—the title, "Chaos in the Brickyard." The bricks (meaning scientific facts) became so abundant the builders could no longer find the right types of brick they needed among the piles, thus the chaos. The final line is "And, saddest of all, sometimes no effort was made even to maintain the distinction between a pile of bricks and a true edifice."

Hopefully you get his point. When science becomes so large, investigators lose track of the larger goals (or narratives), eventually contenting themselves with just gathering facts. This is exactly what Dobzhansky was talking about—gathering sundry piles of facts, many of which are interesting or curious, but ultimately failing to create a meaningful picture.

Journalist David Weinberger conveyed the relevance of this to today's world in a 2012 article in the *Atlantic* titled "To Know but Not Understand: David Weinberger on Science and Big Data." He opens by citing the "Chaos in the Brickyard" essay then says, "If science looked like a chaotic brickyard in 1963, Dr. Forscher would have sat down and wailed if he were shown [today's] Global Biodiversity Information Facility."

The bottom line for all this is that science didn't melt down in the 1960s. It survived the chaos in the brickyard just fine. Similarly science is not facing a crisis today. But it's still tragic to see needless waste, which is what happens when information is gathered without a clear purpose. Filling in the blanks of the Dobzhansky Template will help reduce this sort of wastefulness.

Now it's time to move to a deeper level of narrative structure.

7

Methods: Sentence

The ABT Template

"The ABT is the DNA of story."
   —Park Howell, author

If there's a heart and soul to this book, this is it. In this chapter I present a template for summarizing your story in a single sentence that is as old as Gilgamesh and as fundamental as DNA.

Pitching in an Elevator

Let's begin in familiar territory. If you've taken any sort of communications workshop or training in recent years, you've probably heard about something called the "elevator pitch." The idea is, you're in an elevator, someone important steps in, said person asks you what you do, and now you've got the amount of time between a few floors to explain your entire project in a manner that will be both concise and compelling.

Here's pretty much the current state of knowledge on how to do this effectively: not much. Let me point to three sources to justify this comment. From an Internet search on "elevator pitch," the first link I got was "The 7 Key Components of a Perfect Elevator

Pitch," by Noah Parsons. I'll comment on this in a moment. Second I found the 2012 bestselling book by Daniel Pink, *To Sell Is Human*, with an entire chapter titled "The Elevator Pitch," in which Pink goes through six approaches to creating one. Then I somehow came across *Elevator Pitch Essentials: How to Get Your Point Across in Two Minutes or Less*, by Chris O'Leary. He breaks the process down into nine elements.

Okay, why did I say the state of knowledge is not that much? Do you see the problem here? It's the difference between a shopping list of six or seven or nine pieces of advice versus just one key, core message that will lead to a good elevator pitch.

Communicating a core message is a part of the leadership element to narrative. It's easier to give a list of things to do than to say, "This is the main thing." It's back to that dilemma of "I would have written a shorter letter but didn't have the time."

None of the three elevator pitch lists have a core message. They are essentially groups of disconnected elements. A single, universal tool is much more powerful. This is especially true if the tool not only gives you the structure but also leads you in the direction of narrative thinking. In the end, this is the most essential aspect of grabbing and holding people's attention. It's not easy to boil things down to one key element, but in this section that's what I'm going to enable you to do, by giving you a single tool.

And seriously, O'Leary breaks his book's chapters into "The 9 C's," which are (of course) concise, clear, compelling, credible, conceptual, concrete, consistent, customized, conversational. But he left off the tenth—complicated.

It Is Indeed the DNA of Story

If there is just one piece of knowledge I hope you take away from this book, the ABT is it. There is no competing template for narrative. This isn't "Randy Olson's way of structuring a story." It is THE way. This template tracks directly back to Hegel and Aristotle. There's only one path to them.

As a scientist you can use the ABT multiple ways, starting with describing your research program in a manner that is both concise and compelling. If you craft a one-sentence presentation of your research program with the ABT, you will (a) not bore anyone, (b) not confuse anyone, and (c) activate the narrative part of the brain (remembering back to Hasson's neurocinematics—you'll be doing the Hitchcock thing to some degree). It's that last element that is the real communications powerhouse: the ABT activates the narrative part of the brain.

ABT: The Universal Narrative Template

The ABT is both old and new. Because of its simplicity, most people have the sense that the ABT is something they were taught in grade school. There is a familiarity to it, which is good and gives a head start to learning it. And yet, as far as I can see, no one has formulated it before. I researched it pretty thoroughly in the fall of 2011 when I first pieced it together. I found nothing. Lots of things that were similar but no ABT.

The ABT Template matches the basic three-act structure that I talked about at the start of this book. A story has three parts— beginning, middle and end. A typical story begins with what is called exposition, meaning a laying out of a few facts—basically the setup of the story. The simplest and most common connector for stringing together the setup facts is the agreement word *and*.

So you begin with one or more facts joined together by *and*'s. Then it comes time for the story to start (a story begins when something happens) and for us to enter the middle of the story. This is where the word *but* comes in. *But* is a contradiction word that causes the narrative flow to switch direction. So we lay out a few facts, then we suddenly reverse direction by saying, "but . . ." This establishes a problem, which establishes a source of tension or conflict, and presto, the listener's brain is lit up and we're telling a story. For a murder mystery, for example, we could say, "There is a small town AND there's a happy family, BUT then the father is found dead on the porch of their house . . ."

The word *but* introduces what is often called the "inciting incident," which is where the story begins and is where we transition from the Ordinary World to the Special World I talked about earlier. At this point we've entered the narrative world, and different parts of the brain have become active.

Once we have established this problem (the father is dead), which points to a question (who dunnit?), then we want to head off on our journey in search of the answer to the question, which we do with *therefore*, a consequence word. We now have a story that is set up.

Or we can use it to craft an entire story, as in, "Some people lived in a village AND they lived their lives in terror from a nearby dragon, BUT then one day a knight slayed the dragon, THEREFORE the people of the town lived happily ever after."

The ABT is very flexible. It is universal. It guides you toward being concise and compelling. And it keeps the story moving, which is crucial.

Advancing the Narrative

When I first started presenting the ABT, I did an initial experiment with an audience. At the start of a presentation, my workshop co-instructors Dorie Barton and Brian Palermo and I put up a slide of an Edward Hopper painting then asked for volunteers to tell what they saw in the painting. Off to the side a friend timed their efforts. The volunteers spent an average of just under 30 seconds as they stared at the painting and listed the things they saw but struggled to decide when to stop. Later in the presentation, after explaining the ABT, we put up the same painting and asked for three new volunteers to describe it using the ABT.

They jumped right in, listing a few observations connected with the word *and* to set things up. Then you could feel they knew they needed to get to the word *but*, which they did. And then no sooner than they said that part, you could feel they wanted to move on to *therefore*. Their times averaged only 13 seconds.

That's called "advancing the narrative." It's what audiences desperately desire. They don't want you to make the same point over and over. They want you to keep things moving toward these milestones of establishing a problem and heading in the direction of its solution. That's what the ABT prompts you to do.

With the final volunteers, each description was quick, concise and confident as they said things like, "There are three people in a room AND the sunlight from the window indicates late afternoon, BUT the woman in the middle appears to be questioning the man sitting down, THEREFORE this looks to be an interrogation." They told the story of the painting. Very simple, very concise and very quick. Such is the power of the ABT.

The Land of Boredom: AAA Structure

So what happens if you never get to the word *but*? This happens every day in the real world. People tell "stories" that go on and on and on—never leading to any sort of climax or conclusion—as their audience gets bored.

I first became sensitized to this when I was invited to visit the Centers for Disease Control in Atlanta after the publication of *Don't Be Such a Scientist*. In preparation I spoke with some of their communications folks. One of them told me about their frustrations at times in dealing with the scientists. She said, "We ask the scientists, 'What would you like us to communicate to the public?' They say,

'We want you to tell the story of the CDC.' To which we say, 'Great, what *is* the story of the CDC?' They reply, 'You know, it's all the diseases we cure here, all the drugs we develop, all the awards we win.' To which we say, 'That stuff is all great, but it's not a story—it's just a list of facts.' A story begins when something happens."

That is your basic rule of thumb and is very important. Until something happens, you're not really telling a story. I observed a "storytelling workshop" at a meeting of public health workers for a particular state where the workshop leaders began by asking each participant, "What is the story of your program?" Every single reply consisted of, "Well, we're based here, and we do this, and we do this, and we're this many years old, and . . ." That's not a story. It would have been nice if the workshop leaders had stopped the participants and pointed this out, but they didn't because, as it soon became clear, the leaders didn't really know what a story is.

Similarly, and all too often, scientists stand up in front of audiences showing one graph after another, doing this same thing. They say, "Here's a graph of feeding rate, and here's a graph of digestion rate, and here's a graph of food size, and . . ." But all they are doing is just laying out a pile of sundry facts (as Dobzhansky warned against).

We can label the structure of "And, And, And" as AAA. It is nonnarrative. There is no story being told, just a presentation of facts. And I can assure you it can get boring, which is one of the two worst ways communication can go (the other being confusing, as we will explore shortly).

For Hasson's neurocinematics experiments, the nonnarrative clip of people walking in Washington Square Park could be described as "People are walking AND some have dogs AND some are alone AND the sun is shining AND there are some trees AND . . ." That's an AAA. And that's boring. Which is why the fMRI shows so little brain activity for viewers of that clip.

The narrative clip of a scene from a Hitchcock film would be something like, "There are four men in a room AND they seem calm, BUT one of them pulls out a gun, THEREFORE someone may get shot." That's an ABT, and that's interesting.

The divide may seem simple but don't be fooled. The difference is as profound as it gets in communication. This is why it makes sense to call the ABT the DNA of story. It's as if we're boiling narrative structure down to the level of base pairs and coding.

The Urtext for the ABT

Where did the ABT originate? You remember from part 2 what I said about Gilgamesh, followed by Aristotle, then the five parts of drama, then Hegel in the early 1800s, who formalized the triad of thesis, antithesis, synthesis? Those are the basics, but there's lots more.

Once you absorb the triad structure and begin looking around, you start seeing it everywhere. In workshops I do with the business community, participants talk about the template they use for case studies. They break their stories down into "situation, complication, resolution" as developed by Barbara Minto with her Minto Pyramid Principle.

It's the same triadic structure as the ABT. Situation (we have this AND this AND this . . .), complication (BUT a problem has arisen with this), resolution (THEREFORE we solved it by doing this). The only difference is that *and*, *but*, and *therefore* are shorter and simpler words. Two of them (and, but) are words you hear hundreds to thousands of times a day, which makes them almost invisible. The third one is a little clunky and can often be replaced with simpler words, but it works well while you are becoming accustomed to the structure because it has a constructive, cueing tone to it (much better than the potentially snide "so?").

The ABT crops up all over the place. One of the great scholars of epic literature from the second half of the last century, Albert Bates Lord, developed his own tripartite narrative structure of "Withdrawal, Devastation and Return" (WDR he called it). He claimed to have identified this pattern of narrative structure at least seven times in the *Iliad*.

The ABT is robust and universal, but did its elements really originate with the creators of *South Park*? Nope. In an interview, Parker and Stone mentioned they figured out their Rule of Replacing over their many seasons of writing *South Park* together. So did it originate with them? Not likely.

I think I have discovered who essentially created the framework of the ABT: the legendary screenwriting instructor Frank Daniel. In the opinion of many experts, he was the greatest screenwriting instructor ever. He immigrated to the United States from Czechoslovakia in the 1960s, founded the screenwriting program at Columbia University, was artistic director at the Sundance Institute for a decade, then taught screenwriting at USC, where I was lucky enough to take his script analysis course in 1995, the year before he passed away.

He was amazing—a pioneer in figuring out the structural dynamics of screenplays using his "Sequence Paradigm," which they taught us at USC. At his memorial service, one of his greatest students, cult director David Lynch, gave the eulogy. In an interview Lynch said of him, "No one understood the art of filmmaking as he did."

So here is what I think is the "urtext" (the original source text) for the ABT. In the transcript of a speech Daniel gave in 1986, we find the following:

Monotony is a problem in first drafts. . . . There are several reasons for it. One usually is the fact that the scenes follow in the forbidden pattern: and then, and then, and then. In such a case immediately you have monotony. In a dramatic story the pattern usually for the connecting scenes is: "and then," "but," "therefore," "but," and towards the culmination "meanwhile."

If you don't have this "but" and "therefore" connection between the parts, the story becomes linear, monotonous. . . . Diaries and chronicles are written that way, but not scripts.

So here we go with breaking down these two paragraphs for all the knowledge they contain, just as we did for the Dobzhansky quote.

Monotony

In the first paragraph Daniel pinpoints the bane of bad science talks—the AAA structure we've already discussed. It is the classic situation of a scientist giving a talk that rambles on and on and on. This outcome is basically the result of the refusal or failure to shape the material. As Daniel says, it is the form of first drafts.

Let's consider a basic question: Who will bear the burden of effective communication? Will it be you, the audience member, who is forced to think through all the data presented and shape it into a story of some sort in your mind?

This approach certainly sounds like the ideal way to present science—pure, untouched by the human perspective, just laying out the data for the audience to judge, letting the facts supposedly speak for themselves. It sounds like a good, honest, "inductivist" approach. But not only is this not the way science is done—it is downright dangerous. When information is presented in a way that is devoid of context, not only does this place a huge burden on the listener, it runs the risk of people misinterpreting the research.

The alternative is that the scientist takes on the burden of effective communication. This means the scientist puts in the hours and hours of thought, drafting, testing, honing and shaping the material, eventually ending up with a smooth round cylinder of a presentation, ready to slide into the circular receptors (figuratively speaking) in the brains of audience members.

The result is a presentation of information in a more shaped form, ideally using the ABT. The presentation might consist of a couple of graphs ("Here are data showing this AND this . . ."), then a graph that illustrates a contradiction ("BUT if we look at these data we see something very different from what was expected"), leading to the presentation of the new work ("THEREFORE I began gathering the following data . . ."). This version of structuring the flow of information activates the narrative function in the minds of the audience, bringing all the known benefits of better communication.

The only catch is that it takes a *lot* more effort on the part of the speaker. This prospect prompts the question, how important is it to you to be understood correctly? In chapter 10, I tell about two groups of scientists who were willing to put in this kind of time and effort in working with me. The results were great. But it involved their investing more time into communication than they ever had before. That's the cost.

AAA: Not that there's anything wrong with it . . .

My intention is not to say there is anything whatsoever inherently wrong in the AAA structure. I am always quick to point this out in my talks because I usually look out into the sea of faces in the audience and see numerous grad students cringing and whispering to those next to them, "Oh, no. I just realized the talk I'm giving tomorrow is a complete 'And, And, And' presentation. I'm skipping the party tonight. I've got to redo my talk."

Seriously. I hear this all the time at meetings. At the reception after I give a keynote presentation, students come up and say, "You destroyed my talk I had all set to deliver—thanks a lot!" But it's always a good thing, and the student is laughing about it because, face it, we all want to give good, concise and compelling presentations.

But to repeat, there's nothing wrong with an AAA presentation so long as the data are accurate. It is a structure that is downright adequate most of the time—probably a little boring, but adequate. The only thing we're really talking about here is the push to reach beyond adequate. I want you to reach all the way up to interesting.

However, there is another potential problem with the AAA format, beyond just boredom. There is the risk that people won't view what you're saying the way you want them to because, again, real communication is not as simple as just "letting the facts speak for themselves."

"Truths cannot walk on their own legs." That's what Karlyn Campbell says in *The Rhetorical Act*. "They must be carried by people to other people. They must be explained, defended, and spread through language, argument and appeal." This is a source of irritation for many scientists, but it's the real world.

Hypothetico-Deductive versus Inductive

Ideally, in the doing of good science, you are using the hypothetico-deductive method of science rather than the inductive method. The inductive approach assumes you walk out into nature with a blank slate—absolutely no biases, no preset story in your head—gather data, then after the fact look for interesting patterns.

Inductivism sounds great in theory but almost never happens because we don't want to be wasteful. There are countless essays about how the scientific method is not as robotic as most people presume it to be. In fact, in 1992 Henry H. Bauer wrote a whole book on it titled *Scientific Literacy and the Myth of the Scientific Method*.

Here's the situation. If you have reason to believe birds are all sitting on one side of a tree because of either food or temperature, there's no reason to waste a huge amount of time and resources measuring radioactivity levels within the tree branches, or potassium levels on the ground, or any of the countless other irrelevant variables. At some point you need to make some judgment calls in deciding on the few things you will measure.

What this means is that most scientists concede they tend to use the hypothetico-deductive approach, where you accept from the outset that you're going to employ some less robotic strategies for the sake of efficiency. For starters, you're going to think through all the potential hypotheses that might account for the patterns you see, then quickly, through deduction, you will dismiss the ones you think would be a waste of time to pursue.

The hypothetico-deductive approach is narrative. It is the same as the ABT. With it you are using that albeit imperfectly programmed organ inside your cranium to do the advance work that will make things more efficient and productive for everyone involved. So the scientific method is not the inductivist/AAA process but is instead more like the ABT/hypothetico-deductive pathway.

Which brings us back to the ABT. It is the age-old structure of logic that works best for the masses. It is infinitely powerful. If you have any doubts about its effectiveness, I will now defer to a couple important historical figures who will validate it for you.

Famous Fans of the ABT

ABE LINCOLN

You read that right. Abraham Lincoln was a member of the ABT Club. The Gettysburg Address is an ABT. And that's about all it is because it is incredibly concise and compelling. I have to thank my buddy Park Howell of Arizona State University for spotting this. Park's realization about the Gettysburg Address is a major reason why he proclaimed that the ABT is the DNA of story.

The Gettysburg Address is one of the greatest speeches in American history. So great that famed PBS documentarian Ken Burns made an entire movie to celebrate the 150th anniversary of its delivery. But there's something Burns failed to note in his film. The speech is an ABT.

Just look at it. Could it be more obvious? It's a grand total of three paragraphs. And the three paragraphs pretty much match the three parts of the ABT. (Note: There are at least five versions of this speech. Their differences are small. I'm using the most widely cited version.)

Here's the first paragraph—clearly a string of statements that could be connected by the word *and*, as I've done within brackets:

Four score and seven years ago our fathers brought forth on this continent a new nation, [AND it was] conceived in liberty, and dedicated to the proposition that all men are created equal.

Although the second paragraph doesn't have a *but*, one fits in just fine.

[BUT] Now we are engaged in a great civil war, testing whether that nation, or any nation so conceived and so dedicated, can long endure. We are met on a great battlefield of that war. We have come to dedicate a portion of that field, as a final resting place for those who here gave their lives that that nation might live. It is altogether fitting and proper that we should do this.

Lincoln has now set up the narrative about what "they say" (a country founded 87 years earlier) followed by his "I say" (that the country is now engaged in a civil war), which changes the narrative direction and establishes his problem.

The third paragraph actually begins with the word *but* and continues the *but* section of the ABT. It's not until halfway through that you can finally feel things shift as Lincoln gets to the consequence or "call to action," telling his audience what needs to be done to solve the problem. As you can see, the word *therefore* fits right in.

But, in a larger sense, we can not dedicate, we can not consecrate, we can not hallow this ground. The brave men, living and dead, who struggled here, have consecrated it, far above our poor power to add or detract. The world will little note, nor long remember what we say here, but it can never forget what they did here. [THEREFORE] It is for us the living, rather, to be dedicated here to the unfinished work which they who fought here have thus far so nobly advanced. It is rather for us to be here dedicated to the great task remaining before us—that from these honored dead we take increased devotion to that cause for which they gave the last full measure of devotion—that we here highly resolve that these dead shall not have died in vain—that this nation, under God, shall have a new birth of freedom—and that government of the people, by the people, for the people, shall not perish from the earth.

So there we are. It's the same old structure. (And by the way, I guarantee you this is a great way to teach students about this speech—get them to view it as an argument with this structure to it, then have them read *They Say, I Say* for more detail on the power of argumentation in learning.) It's pretty stunning when you look at it—the tripartite structure, right there, clear as day, hiding in plain sight.

Abe made his point, got on his horse and rode away. Not surprisingly the speech lives on. Yes, the speech has lots of other fine attributes—a great opening with a wonderful cadence "Four score and seven years ago" (it is often pointed out how much more lyrical that is than just "Eighty-seven years ago . . ."), brevity, simplicity (I'm drawing these from a source that addresses the question of why it is such a great speech)—but it could have had all those and still been mediocre. An AAA statement can be brief and simple yet still forgettable.

One source I found praises the speech for having a tripartite structure but views the structure as "past, present, future." Which it is. But if the present part had not been problem oriented, it wouldn't have been powerful.

It's the problem-solution thing. It immediately puts you into the narrative realm, activating that part of your brain. The bottom line is that the speech is a paragon of ABTness.

WATSON AND CRICK

Here's another example of the ABT structure in a famous work. If you're a scientist, this is hopefully the clincher for you. In 1953 James Watson and Francis Crick published in *Nature* what is probably the single most important research paper in biology if not in science in general. It was their initial description of the structure of the central building block of life, Deoxyribonucleic Acid, DNA.

Their paper is legendary for being not just compelling but also extremely concise, with a length of a mere two pages. James Watson in later years showed how gifted a writer and teller of stories he is with the publication of his book *The Double Helix* in 1968. He clearly has a deep feel for narrative. And not surprisingly, the *Nature* paper falls right into the ABT structure on the first page.

Look how it starts—with a series of expositional statements that could be strung together with *and*'s. The authors basically say (paraphrasing here): a structure has already been proposed by other folks AND those folks made it available to us AND their model has three chains BUT . . . we think they are wrong.

So simple, so clean, so eloquent. It is their argument, laid out plainly. Straight out of *They Say, I Say*. They lay out what "they say" (Pauling and Corey) then transit into their "I say" segment ("In our opinion this structure is unsatisfactory . . .") then present their new findings.

They took their time bringing you into their world narratively. They didn't rush things by heading off in multiple directions. They

BUT

A Structure for Deoxyribose Nucleic Acid

We wish to suggest a structure for the salt of deoxyribose nucleic acid (D.N.A.). This structure has novel features which are of considerable biological interest. A structure for nucleic acid has already been proposed by Pauling and Corey. They kindly made their manuscript available to us in advance of publication. Their model consists of three intertwined chains, with the phosphates near the fibre axis, and the bases on the outside. In our opinion, this structure is unsatisfactory for two reasons: (1) We believe that the material which gives the X-ray diagrams is the salt, not the free acid. Without the acidic hydrogen atoms it is not clear what forces would hold the structure together, especially as the negatively charged phosphates near the axis will repel each other. (2) Some of the van der Waals distances appear to be too small.

Another three-chain structure has also been suggested by Fraser (in the press). In his model the phosphates are on the outside and the bases on the inside, linked together by hydrogen bonds. This structure as described is rather ill-defined, and for this reason we shall not comment on it.

We wish to put forward a radically different structure for the salt of deoxyribose nucleic acid. This structure has two helical chains each coiled round the same axis (see diagram). We have made the usual chemical

THEREFORE

Figure 12. Watson and Crick used the ABT form in their landmark 1953 *Nature* paper. It opens with the ABT structure, which you can see if you insert the ABT words.

just gave you a clear idea of what their work is about then took you in a single new direction.

They also didn't go on for too long. They didn't cite 15 opening facts to impress you with how much they know, pushing you to the point of wanting to say, "Okay, I get it." They set up their "story" then started it.

That's how the Greeks did it. It's how George Lucas did it in *Star Wars*. It's how most people, both technical and nontechnical, would prefer you do it. First get us oriented, then point us in *one* new direction. This is a timeless dynamic that was around long before the first scientific journal.

Word Dynamics

Now let's take a closer look at the three words that make up the ABT. They represent three types of "transition words," which match the dynamics of three-act structure. They go from agreement to contradiction to consequence.

AND

And is a word of agreement and positivity. If you know much about improv training, you know that it's all based around the idea of affirmation, which is achieved by saying "yes" to any suggestion followed by this word of agreement. The standard improv catchphrase is "Yes, and . . ." If someone says something preposterous to you, you don't negate what they've said, you affirm it with "Yes, and . . ."

By connecting your opening facts with *and*'s, you are starting your story/argument/explanation without tension or conflict—just pure agreement, just laying out some basic facts before you start to challenge the minds of your audience. Appendix 1 includes a list of other agreement words such as *also, likewise, similarly, as well as, in addition.*

BUT

But is a word of contradiction, negation and denial. Most improv instructors forbid the use of this word. It changes the direction things are headed, which is a bad thing if you're trying to be creative and build big ideas. It kills creativity as it changes the direction of flow.

What occurs with *but*, because of the contradictory direction it forces, is the establishment of tension, or even conflict. We were happy going one direction. We were comfortable. BUT . . . now we're not going that direction any more, which makes us uncomfortable.

Conflict is the driving force of all stories. Robert McKee, in his book *Story: Style, Structure, Substance, and the Principles of Screenwriting,* describes what he calls his "Law of Conflict for Storytelling." He says, "Nothing moves forward in a story except through conflict." He adds, "Conflict is to storytelling what sound is to music."

This is what we feel when Watson and Crick say in their *Nature* paper, "In our opinion . . ." They have laid out the introductory facts—what others "say"—but then they change the direction. Thus the story moves forward, the narrative is advanced, and everyone is drawn in deeper, instead of losing interest.

Here are some other contradiction words: *despite, yet, however, instead, conversely, rather, otherwise*. What's important to note is that sending the narrative off in a different direction is a great thing, once. But if you think about the power of the singular narrative, you can see that more than once can be a problem, as we will discuss soon.

THEREFORE

Therefore is a word of consequence. It is a "time word." It shows up after some amount of time and signals a consequence or effect.

What is the central element in a story? Time. When we talk about advancing the narrative, we're talking about moving things forward in time. That's what *therefore* does. It pulls things together and moves them further along.

I actually observe *therefore* functioning like this in my workshops. Someone will be trying to make a point but end up going on and on at length until finally someone else, almost involuntarily, blurts out, "Therefore . . . ?" It becomes a cue, meaning "What's your point? What are you getting at? Where are you going with this?"

Recently I was at a very, very boring talk on climate change, seated next to an actor buddy of mine. I had completely forgotten that he once attended one of my talks. So I was delightfully stunned when, after a half-hour of listening to the droning speaker, he whispered to me, "Therefore . . . ?"

The ABT Words as Scaffolding

There's nothing sacred about the three ABT words. You don't have to use those exact choices. You could use "Also, Still, Since" instead (though you'd end up with a rather unfortunate acronym). But what's even better is not using them at all. They are simply building elements that help you get to the ideal narrative structure. Then, if you've built a strong edifice with them, they can be removed and the building will still stand up on its own.

This is why you don't see them in the Gettysburg Address or the Watson and Crick paper. They aren't needed, but the fact that they work when you add them in shows you the writers had good narrative intuition.

In fact, the ABT can be used as a test—a sort of null hypothesis for whether a given piece of writing has good narrative structure. And this is exactly where we are headed. But first, we need to add one more configuration.

DHY: Major Confusion Meets General Boredom

Okay, the title of this section is a tribute to my all-time favorite movie review. It was for the Civil War movie *Gods and Generals*. Some heartless reviewer titled his review "Major Tedium Meets General Boredom."

There are two significant ways communication goes bad. The first is boredom. The second is confusion. The AAA produces boredom. What we're going to talk about now is the configuration that leads to confusion.

At this point we know too well that if you have too many *and*'s you end up with the AAA structure, which is nonnarrative. And we have learned that the ABT structure can produce a Goldilocks narrative ("just right"). Now let's add a third category, which I'm going to describe as "overly narrative" (I would describe it as "hypernarrative," but I see that word in wide use in the world of gaming, and I'm not sure this matches their concept.)

What I mean by "overly narrative" is the presence of too many narrative directions, delivered either all at once or sequentially. It's basically a story that's too confusing to follow. I'm sure you've heard at least one in your lifetime.

It can happen through the use of too many contradiction words. Having a single contradiction word such as *but* establishes one source of tension and conforms to the power of one we have talked about as giving us a nice, clean, singular narrative. But if you look around, you'll find plenty of instances of where the storyteller is winding out numerous narrative directions all at once or leading the audience down a zigzagging path.

For shorthand I'm going to abbreviate this mode using the three contradiction words of Despite, However, Yet (DHY). (In chapter 10, I will point you to the abstract of a research paper that actually presents three consecutive sentences that start with those three words.) You can see what the result is—*despite* sends you in one direction, *however* sends you in another, and *yet* creates one more direction. All of which adds up to confusion for most people.

This can be a more interesting, more challenging, and to some people more rewarding way to communicate than ABT. If you pull together a group of people who think in this multinarrative, convoluted way, they will probably have a great time together.

You hear this mode of narrative all the time among academics. They thrive on complex narratives in the same way that they love to speak at times in a sort of double negative way, saying things like, "There were a nontrivial number of people, and let's just say what they were doing was not unimportant." You can talk that way and some people will enjoy it. But not the masses. They don't talk like that. They say, "There were a significant number of people, and they were doing something important."

For the science world, Anne Greene addresses this need for using "plain language" very nicely with her book *Writing Science in Plain English*. She says, "Scientific writing, while exploding in quantity, is not improving in quality." She seeks the same element of simplicity in the use of words that I'm pushing for with narrative.

The DHY structure—heading off in multiple narrative directions—makes me think of my directing teacher in film school, Eddie Dmytryk. He was one of the founders of the entire genre of film noir in the 1940s. He showed us some of his noir films—*Crossfire, Murder My Sweet, Cornered*. He warned us that the plotting was *very* complex, which was an understatement. I was lost within the first 10 minutes of each film. I'd find myself thinking, "Wait, did she kill that guy? I thought she was his wife—no, hang on, that was her sister. But wasn't her sister already dead? I'm lost."

Dmytryk compared them to the *New York Times* advanced crossword puzzle—the sort of puzzle where the clues have secondary and tertiary levels of meaning that only the superstar solvers perceive.

This then becomes the question for scientists: who are you wanting to talk to? If it's just the seven people in your field who know all the material at the same depth as you, then fine, go ahead with the DHY and have your narrative head off in five directions at once. But if it's the scientific community in general, or the general public, you should look to Watson and Crick as your role models. They are the proof of what Da Vinci had to say about simplicity being the ultimate sophistication.

The Narrative Spectrum

So now we have three variations on narrative structure—none (AAA), optimal (ABT) and too much (DHY). This takes us back to figure 2 in the introduction. We've now assigned a narrative structure to the idea of being boring, interesting or confusing. I'm calling this the "Narrative Spectrum."

I think this may prove to be a powerful tool. With it you should be able to look at any bit of writing or speaking and place it somewhere along the spectrum. If the writer or speaker is losing the audience, they are probably either boring them (with a nonnarrative approach) or confusing them (with an overly narrative approach). Or both.

This is at least an analytical way to look at failed communication and get some idea of where to improve things. It is more analytical than just saying, "I don't know, he lost me." Analytical is good—especially if you're a scientist. In chapter 10, we will put the Narrative Spectrum to work.

Is This Just for a Past Generation?

It's time for yet another clincher. Maybe you're thinking Watson and Crick were writing in 1953; Abe Lincoln, a century earlier. Perhaps you think times have changed, that today scientists need a wider range of narrative structures and shouldn't get locked into the ABT Template.

Wrong.

In the introduction, I mentioned 2013 Nobel laureate Randy Schekman's call for a boycott of the top scientific journals. In return for popping his head up like that, let's just use him as a target for this Narrative Spectrum test. Does he show good narrative structure in the abstracts of his research papers?

Guess what—in looking at the abstracts of his six most recent papers, I found that four of them have solid ABT structure and the other two are close. Even though they are all written with so much discipline-specific terminology that I can't begin to tell you what they are about, the structure shows through.

Here's one of them. See if you don't agree that it has clear ABT structure. The paper is titled "Regulated Oligomerization Induces Uptake of a Membrane Protein into COPII Vesicles Independent of

Its Cytosolic Tail." I have no clue what all that means, but here's the abstract:

Export of transmembrane proteins from the endoplasmic reticulum (ER) is driven by directed incorporation into coat protein complex II (COPII)-coated vesicles. The sorting of some cargo proteins into COPII vesicles was shown to be mediated by specific interactions between transmembrane and COPII-coat-forming proteins. But even though some signals for ER exit have been identified on the cytosolic domains of membrane proteins, the general signaling and sorting mechanisms of ER export are still poorly understood. To investigate the role of cargo protein oligomer formation in the export process, we have created a transmembrane fusion protein that—owing to its FK506-binding protein domains—can be oligomerized in isolated membranes by addition of a small-molecule dimerizer. Packaging of the fusion protein into COPII vesicles is strongly enhanced in the presence of the dimerizer, demonstrating that the oligomeric state is an ER export signal for this membrane protein. Surprisingly, the cytosolic tail is not required for this oligomerization-dependent effect on protein sorting. Thus, an alternative mechanism, such as membrane bending, must account for ER export of the fusion protein.

It is an object of ABT beauty. Schekman and his coauthors open with two clear statements of exposition, and then look at what word we see—*but*. They state the problem, which is that something is "still poorly understood." And then you could insert a *therefore* to transition into "To investigate the role of . . ."

Here is the first part again, with my editorial comments to show you the ABT structure.

STATEMENT: Export of transmembrane proteins from the endoplasmic reticulum (ER) is driven by directed incorporation into coat protein complex II (COPII)-coated vesicles. STATEMENT: The sorting of some cargo proteins into COPII vesicles was shown to be mediated by specific interactions between transmembrane and COPII-coat-forming proteins. CONTRADICTION: But even though some signals for ER exit have been identified on the cytosolic domains of membrane proteins, the general signaling and sorting mechanisms of ER export are still poorly understood. CONSEQUENCE: To investigate the role of cargo protein oligomer formation . . .

Pretty much, ipso facto, case closed if you ask me. The other abstracts are just as cleanly written. And guess what the guy won—the Nobel. Thank you Randy Schekman for giving me such a great set of samples, even if they might as well be in Chinese in terms of my ability to understand the content. It's the form that I can spot as clear as day.

All We Are Saaaaaying Is Give Simplicity a Chance

All we are seeing here with this ABT structure is a finer-scale version of what the science world began realizing a century ago at the larger scale with the IMRAD template. If you're wanting to object to the idea of conforming to the ABT Template, you should ask yourself if you feel like objecting to the IMRAD template as well.

For a century now the science world has stifled expressive creativity in research papers at the larger scale by mercilessly imposing the IMRAD template on all writers. To which almost all scientists would reply, "Thank goodness," when they think about what it would be like to read papers without this structure.

The reason behind this is the same as for what I'm talking about. The main purpose of scientific writing is not to showcase wild and expressive creativity. It is to convey important and interesting information in a manner that is maximally efficient and minimally misunderstood. All I'm suggesting with the ABT is taking the IMRAD approach down another level in the interest of minimizing boredom and confusion. You can start with the abstract, but you really want to use it throughout. Remember, Always Be Telling stories.

Shaping: Concise versus Compelling

Hopefully I've now convinced you that the ABT is a powerful narrative tool worth using to structure or present your own stories or research projects. So if you've decided that, yes, you'd like your research to have the clarity and breadth of Watson and Crick, then let me now guide you toward crafting your own ABT that is both concise and compelling.

First, think of how those two properties work in opposition to each other. The desire to be concise forces you to chop your content way down. It's mandatory in today's information overloaded world. But of course if you go too far, you end up with the problem of "dumbing down," which is not what you want (though I get accused of it all the time from people who are not interested in listening to what I'm saying here).

So you want to trim your content way down yet still retain the important pieces of information. What I'm about to present here is an approach that is analytical, not just intuitive. I mentioned a while back several authors urging you to develop an elevator pitch, and I pointed out how almost all of their suggestions are just vague, intuitive advice to make the pitch short, punchy, lively, yet retain the "essence" of your work. None of it is analytical. None of it helps you with the actual mechanics. But the ABT does, especially with this procedure.

The Goldilocks Approach to the ABT

We're again going to do the Goldilocks thing by creating three ABTs—one that's too big, one that's too small, then finally one that's juuuust right. We'll do this by varying the two parameters of being concise and compelling.

This exercise helps you understand what you are trying to say with the story you're wanting to tell. This is a problem that comes up all the time in my workshops—people with a big story to tell that is all jumbled up. I ask them, "What is the story you want to tell? What do you want us to know, exactly?"

1. THE INFORMATIONAL ABT (iABT)

It may seem a little overstated, but I'm going to label these three versions of the ABT with lower case letters. It feels like I'm trying to emulate messenger RNA (mRNA), transfer RNA (tRNA), mitochondrial DNA (mtDNA) and other incredibly important components of the beginnings of life itself. Actually, why not? As I've said, the ABT is the DNA of story.

Our starting point is the Informational ABT (iABT). This is a first version where you don't worry at all about being concise. Our only interest is to include all the compelling information.

What this produces is a massively long, clunky "sentence" that you hopefully would never try to speak in public. It's just the whole enchilada as a starting point, meant to include all the potentially compelling and interesting information, yet still narratively structured using the ABT words.

As an example here's an iABT from Katelynn Faulk, a graduate student from the University of North Texas who took part in my workshop at the American Physiological Society meeting in 2014:

iABT: In my lab we model moderate sleep apnea in rats with a chronic intermittent hypoxia protocol in order to investigate the physiological mechanisms of sustained diurnal blood pressure, BUT we have realized the importance of molecular pathways within the central nervous system contributing towards blood pressure control, THEREFORE we have begun exploring novel molecular pathways that develop as a result of our sleep apnea model.

Okay, whew, that's a mouthful. Katelynn wouldn't want to be caught dead saying all that at a cocktail party if someone asked her what she does. But don't worry. It's just a starting point.

2. THE CONVERSATIONAL ABT (cABT)

Now we go to the other end of the range, creating an ABT that takes concision to its extreme. The Conversational ABT (cABT) is a much more interesting configuration of the ABT for a couple of reasons. First, it reveals the core argument being made, and second, it provides the chance for "narrative relatability," which I will explain shortly.

The first challenge in creating the cABT is to strip the sentence of all of the compelling information and context. Put the ABT into the most generic form possible. I know this is going to read as funny, but trust me, this is what you want to uncover beneath all the words in the iABT.

Here's what I helped Katelynn arrive at (I say this so you can blame me if you think it sounds pretty dumb):

cABT: We were looking at one way but realized there's another way therefore we're looking at that way.

Yep. Sounds pretty dumb. But it's what we want—and it's actually not dumb, just totally generic and free of context.

The first thing you gain from this exercise is the realization of what it is, at the very core, you are saying. This is your story in its simplest form. This is what you can say when someone asks you, "What exactly are you trying to say?" You must be sure to always have an answer to that question—which, too often, participants in my workshop do not. Here Katelynn can answer, "Basically we were doing things one way but found out there's a better way to do it, which is what we're working on now." That is the core of the "story" she wants to tell.

Now let's pause our Goldilocks ABT discussion to talk about why this is a powerful element.

CHARACTER RELATABILITY VERSUS
NARRATIVE RELATABILITY

In our Connection Storymaker workshops, the idea that our improv instructor, Brian Palermo, advocates most is the need for "relatability." At the start of our book *Connection*, each of us offered up a one-sentence summary of our main message. Brian's sentence is "Make your story relatable."

This is yet another aspect of taking the communications burden on yourself, which takes time and energy but is important. Brian is saying that if all you do is tell people a bunch of facts about your life, they may or may not find it interesting. But ultimately, they're probably going to wonder, "So what does this have to do with me?" because they simply can't relate to what you're saying.

Brian recommends finding some way to shape what you have to say into a form that your audience can relate to. If you're speaking to a group of golfers about the physics of space flight, see if you can present some of the challenges in terms of the physics of golf. Anything you can insert that they will recognize from their world will make it easier for them to relate to what you have to say.

We can call that "character relatability"—using character material that bears direct similarity to their world. This is powerful and important. But by following the ABT process, it's also possible to connect through what we can call "narrative relatability." This is a new distinction I've begun to make in our workshop with the ABT. I don't see anyone talking about narrative this way in the books on story structure published so far, but I think it has the potential to be powerful.

Say you are speaking to a group of people who have absolutely no background or interest in your field. You might still connect with some of them for at least a moment if you have a narrative structure they can recognize and relate to.

Suppose you begin by saying, "Let me tell you what I've been up to in my lab lately. We've been doing things one way but recently realized there's another way to do them, so now we're looking into that."

It's entirely possible that one person in the group is a realtor and is suddenly thinking to herself, "Wow, that's just like me—I've been using one listing service for years but recently found out about a new one and now am trying the new one."

For that one instant that person will be thinking she's got something in common with you. You will have opened a channel of communication by offering up something relatable.

Now if you go on to say, "It all began when my new assistant offered up a suggestion," it's entirely possible that the realtor will think, "Wow—that's also how it began for me. I hired a new assistant and he told me about this other listing service." Once that happens, she is going to track you down later, tell you about how much you have in common, and you'll be buddies for life.

But in contrast, if you started your talk by saying, "Let me tell you about the chronic intermittent hypoxia protocol I've been using in my laboratory . . . ," the realtor as well as all the other nonscientists will instantly disconnect. Your communication possibilities will be over.

Keep in mind that the relatability has to come first. A woman told me about a dinner she attended in Australia where she sat across from the CEO of a big mining company. She immediately began lecturing him about global warming and he shut down. She asked how she might have taken better advantage of the opportunity.

I told her she could have begun with some character relatability. If she had been able to Google the guy and find out he was, say, an avid tennis player, she could have begun by talking about her favorite tennis players. Basically she just needed something, anything, to provide common ground and open up the channels of communication.

But you have to lead with the relatable material. It's not going to work to get into a spat about environmental practices with the guy, then try to change the subject by saying, "So, did you happen to catch the Australian Open?" Nope. That won't work at all.

3. THE KEEPER ABT (kABT)

The Keeper ABT (kABT) is your finished product. The length will be somewhere in between the other two ABT versions. You get to it by adding back some of the information you stripped out, bit by bit, while maintaining a balance between retaining concision and making it compelling. The cABT was too vacuous to be of use in presenting your story publicly, but you don't want to slip back to something as clumsy and huge as the iABT.

This was my suggestion for Katelynn's kABT:

kABT: In my lab we're studying sleep apnea using rats as our model system, AND we've been focused on physiological mechanisms, BUT lately we've realized the real controls may lie at the molecular level in the central nervous system, so AS A RESULT we've begun exploring novel molecular pathways.

This version is short enough to roll off her tongue yet includes compelling pieces of information that tell her basic story. This is roughly what she'll want to say when that VIP in the elevator asks, "So what sort of research do you do?" Her reply: "Well, thanks for asking. I study sleep apnea. Yeah, I know, kind of wild. In my lab we actually use rats as a model system and for a while we've been focused on physiological mechanisms as the controls, but *recently* we've realized the real answers are probably at the molecular level in the central nervous system, so now we're changing directions and looking at molecular pathways. And that's my story—a shift from physiological to molecular levels."

It's simple, clear. It moves right along toward an overall point. It's the sort of statement that won't bore or confuse. In fact, for many it will arouse their interest. Such is the power of narrative.

One point more. Sometimes people ask, "How do I know how many words an ABT should be?" My answer is simple—intuition. There is no set length. It will be different for every story. You will probably even want to come up with more than one ABT for whatever project you're presenting, as well as different ABTs for different audiences. You'll want one that is light on the jargon for the broadest audience, but then one for your colleagues that has a little more technical language.

But when it comes to the length, that's where you need to have the narrative intuition that is the goal of all of this. That is your only hope for the long term—to be able to just feel how many words you need rather than working toward a set number, because there is no set number.

8

Methods: Paragraph

The Hero's Journey

It's time for the third tool, the Paragraph Template, which is big and fun and the heart of storytelling, and yet . . . will take you a long time to fully grasp and utilize. For this reason, I opted to spend much more time on the Sentence Template (the ABT), which is more practical at the beginning stage of this journey. Here I simply offer a glimpse of where the road can take you, eventually.

The Paragraph Template is built around the real Joseph Campbell material, the Hero's Journey. It's the stuff George Lucas used for creating *Star Wars*. It's fun to fiddle around with it so you can impress your friends—"Hey, I'm tapping into the power of Hollywood for my communications"—but be careful at this level as it can definitely lead you astray. It is easy to misuse, misunderstand and then get frustrated with. I've seen it in our workshops. It happens.

I've already told you about Joseph Campbell's circle diagram for a story. The Hero's Journey is the more detailed version of it. The best thing I can do for you at the outset is refer you to a few excellent resources.

Hooray for Campbellwood

Christopher Vogler's *The Writer's Journey* is the definitive resource for the power of drawing on Joseph Campbell for telling stories. You've also seen how widely I've cited Robert McKee's *Story*. For the practical application of that material—not in story writing but in the business world, there is the excellent 2007 book *Winning the Story Wars: Why Those Who Tell—and Live—the Best Stories Will Rule the Future*, by Jonah Sachs. Lastly, Matthew Winkler's *What Makes a Hero?* is a great, simple, short TED-ed video you should view— especially the first two minutes. It will give you a good overall intuitive feel for how Campbell's circular model of the Hero's Journey works.

As I've mentioned, the pivotal event in Hollywood for the realization and appreciation of story structure came when George Lucas applied Joseph Campbell's teachings to the first *Star Wars* movie. Nothing was ever the same after that. Many other admirers and even worshippers of story structure came along afterward, further spreading the gospel of Campbell. Which was fine for a while, but nowadays concerns have arisen.

One of those Campbell prophets was Blake Snyder, who wrote the 2005 book *Save the Cat: The Last Book on Screenwriting You'll Ever Need*. It was almost an instruction manual for screenwriters on how to assemble a story. The book became hugely popular, but there has since been backlash and serious concern that the industry of moviemaking has become overly formulaic.

The best recent essay on this concern is the July 2013 *Slate* article by Peter Suderman titled "Save the Movie! The 2005 Book That's Taken over Hollywood and Made Every Movie Feel the Same." The 2005 book he is referring to is, of course, *Save the Cat*. Suderman's essay is worth reading. He summarizes the basic problem of movies these days—that they feel so similar to each other. He says, "Summer movies are often described as formulaic. But what few people know is that there is *actually a formula*—one that lays out, on a page-by-page basis, exactly 'what should happen when' in a screenplay. It's as if a mad scientist has discovered a secret process for making a perfect, or at least perfectly conventional, summer blockbuster."

Suderman's concern and complaint that recent movies all feel the same is valid. In fact, what's so great about his article is that he used the very Hero's Journey formula he's talking about to structure the essay itself. At the end he reveals this and invites the reader to review the text and spot the various moments. Sure enough, he hits the "darkest hour" halfway through his essay as he says, "Once you know the formula, the seams begin to show. Movies all start to seem the same, and many scenes start to feel forced and arbitrary, like screenplay Mad Libs." He finishes with these conclusions about the Hero's Journey formula:

> It helped me order my thoughts and figure out what I should say next. But I also found myself writing to fit the needs of the formula rather than the good of the essay—some sections were cut short, others deleted entirely, and other bits included mostly to hit the beat sheet's marks. It made writing easier, in other words, but it also made me less creative.

Okay, let me say a couple of things in response to these very true words.

Hollywood Is a Cesspool of Noncreativity

I can hurl this criticism because I saw it happen, from the very start of film school—basically lazy writers recycling the same movie material.

I showed up at film school wanting to make movies about my 20 years of science experience in the real world. But the majority of my classmates wanted to make movies using the same elements of all their favorite movies, from *Indiana Jones* to *Die Hard*. They didn't want to bring in fresh, new, different material from the real world. They actively and eagerly wanted to recycle (which by then Quentin Tarantino had shown could be incredibly hip and cool, if sometimes stagnant).

In our writing classes I watched these students create characters for their screenplays that were obviously based on characters they knew—not from the real world, but from the countless movies and TV shows they had consumed in their youth. They were like Jim Carrey in *The Cable Guy*, whose character's entire persona and perception of life is fabricated from the various sitcom characters he grew up with. There simply wasn't then, nor is there now, any ethic for or interest in reaching out for new material. It is creative collapse and no one has a problem with it.

How the material is shaped and whether the shaping is all the same is kind of trivial. The real problem is that of endless recycling. Or as a writer friend likes to put it, "They like to breathe their own exhaust."

The Problem Is Content, Not Form

When all the houses on a street are built of the same bricks in the same form, that form is instantly noticeable. But when they are built of different materials in the same form, that form can actually be unnoticeable, at least for a while. And more importantly, it can even be enjoyable if there are elements of creativity added to the selection and fine-scale attributes of the materials.

Again, this is the real problem of Hollywood. It's the source material, not the form, that produces boredom. And just to underscore this, I'll tell you about a blog post a friend sent me a couple years ago about a screenwriter being told what the studios mean when they say they are seeking "originality."

The writer described a pitch meeting with a group of producers who kept saying they were looking for "original" material, which was what the writer thought he was presenting, but they didn't like his stuff. One of the producers finally said that what they meant by "original" was something like a popular, bestselling comic book character that has *never* been made into a movie but is already hugely popular in the comic book world. That was their definition of "original."

In that world, a character you have come up with in your mind from scratch while slaving over your screenplay in your apartment, well, that's not what producers mean by original. That's more like just weird because nobody's ever heard of the character. And they don't want to get involved with weird. Seriously. This is the mindset of a lot of Hollywood producers. Familiarity is their life's blood. Different is just kind of, you know, weird.

And yet they are shocked when audiences say the movies all feel the same. But again, it isn't the form that's creating this feeling. It's the limited variation of the content.

Two Variations of the Paragraph Template: Logline Maker and Story Cycle

LOGLINE MAKER

For a Hollywood movie, the entire story is usually distilled down to a single sentence or paragraph called a "logline." In our book *Connection* my coauthor Dorie Barton created a template she called the "Logline Maker," which is based on Blake Snyder's *Save the Cat* structural elements (which, in turn, are based on Joseph Campbell's Hero's Journey model):

1. In an ordinary world
2. A flawed protagonist

3. Has a catalytic event that upends his/her world

4. After taking stock

5. The protagonist commits to action

6. But when the stakes get raised

7. The protagonist must learn the lesson

8. In order to stop the antagonist

9. To achieve his/her goal

This Logline Maker is another fill-in-the-blanks template. We built this tool into our Connection Storymaker app, which can be a lot of fun at a party as everyone shouts out the elements—"Okay, somebody tell me an ordinary world . . ." "A meat-packing plant!" "A shoe repair shop!" "A nail salon!" "Okay, now a flawed protagonist . . ." "A kleptomaniac jeweler!" "A dishonest cop!" "An alcoholic priest!"

On and on as all nine elements are filled in, Mad Libs fashion. But we realized early on, "garbage in, garbage out." It can be fun having everyone shout out the suggestions, but the paragraph you end up with—which is usually the funniest thing of all—will probably be useless nonsense.

If you do that exercise too many times, you begin to believe the whole idea of the logline is useless and silly. But if you approach it seriously, what you get back will not be silly. It may take you a while to get it to a functional form, but eventually it will be useful.

STORY CYCLE

I'm not going to spend time on the Story Cycle—the alternate to Dorie Barton's Logline Maker and other Hero's Journey templates. Many books have now been written on the Story Cycle and continue to be. But for comparison with the Logline Maker, here are the 12 standard elements of the Story Cycle based on the version presented in Matthew Winkler's TED-ed talk:

1. Call to adventure
2. Assistance
3. Departure
4. Trials
5. Approach
6. Crisis
7. Treasure
8. Result
9. Return
10. New life
11. Resolution
12. Status quo (but upgraded)

Strengths of the Hero's Journey Model

Remember figure 11, the graph of return over time for the three WSP templates? The Word and Sentence Templates have immediate use, but it's the Paragraph Template that will eventually take you the full distance to achieving narrative intuition. It's tempting to sit down and try to put it to work right away for your own research program, but you have to be careful. It's easy to end up with a mess.

Here's how quickly things can go off the rails. Using the Logline Maker, you might say, "Okay, we're going to have our research laboratory be the protagonist—so what's our flaw? Let's say it's that we're never on schedule. Okay, what's the catalytic event? . . ." And onward, filling in the blanks. But until you've had a lot of practice with this template, you'll probably hit blanks you can't quite fill out. You'll end up saying, "Wait, who or what is our antagonist? Is it our funding agency? Or is it the public? Or is it just the time schedule we're up against?"

This confusion is the very thing Suderman is complaining about when he says, "I also found myself writing to fit the needs of the formula rather than the good of the essay." That's a mistake you don't want to make. If you are smashing things into a template even though they don't seem to fit, you've probably jumped the track. This is exactly where you risk "bending the science to tell a better story"—which is what a *New Scientist* book review critic accused me of advocating in *Don't Be Such a Scientist* (even though I wasn't, as I was allowed to point out in a rebuttal letter to the editor that they published two weeks later). You never want to bend the science.

So be careful not to force the Hero's Journey Model to work for you. And that is my warning about the proximate value of this tool—it may not have any value to you whatsoever in the short term. Don't try to squash your story into it.

It's in the long term where the Hero's Journey Model becomes truly valuable. The more you work with it, the more familiar you become with it, the more you absorb it to the level of second nature, the better and stronger you become with the use of narrative. If you get to know it well, you'll end up spotting the elements, by themselves, in real world situations.

For example, here are four specific things I've spotted within the Hero's Journey model (using the two variations of the Paragraph Template described above) that help with developing a deeper sense of story structure. There are lots more than just these.

1. THREE PROBLEM-SOLUTION COMBINATIONS

As I've pointed out repeatedly, stories, at their core, are about problems posed and solved. The Logline Maker, within its structure, presents three problem-solution situations.

The first can be seen in elements 3 and 5. A problem arises in 3 when the Ordinary World of the protagonist is upended. The solution (at least the initial, temporary solution) is presented in 5, when action is taken.

The second problem-solution combination happens in 6—when the stakes get raised, meaning that a new problem has been presented. The solution to this one is sort of mixed between 7 and 8 as the protagonist figures out how to overcome his or her flaw.

And that's the third overall problem-solution combination—the flawed protagonist. In 2 the problem is presented in the flaw of the protagonist. In 7 the protagonist must solve the flaw.

When you think about this, you begin to realize why this is the blueprint for a really powerful story. Think about how problem-solution oriented we are as creatures to start with, then think about a triple dose of the problem-solution dynamic.

You might respond to that comment by saying, "I thought you wanted only a singular narrative?"—which is true. This still is. It is still just a single protagonist, and the singular problem is what arises in 3. The whole story is about the one character addressing that one problem. It's just that the stakes (for the same problem) get raised, and the problem is ultimately solved by addressing the protagonist's flaw.

2. THE FLAWED PROTAGONIST

Now think about the Logline Maker's flawed protagonist element in terms of the science world. We have talked about the tendency for people to believe that the scientific method is a pure, robotic process carried out by flawless individuals who eventually find their way to the truth. Despite the endless string of essays and books written trying to point out that no, science and scientists are not flawless (such as Henry Bauer's *Scientific Literacy and the Myth of the Scientific Method*, which I mentioned earlier), there lingers this deep feeling that scientists are, and should be, flawless robots. Think of Spock from *Star Trek*, who was essentially the embodiment of the supposedly perfect scientist type. He was constantly marveling at how flawed humans are in their basic reasoning. They could never be him.

So here's the dilemma that the Logline Maker shows you: the public loves and needs their heroes to be flawed. Oskar Schindler had to overcome his greed in *Schindler's List*. Rocky Balboa had to overcome his self-perception as a loser in *Rocky*. Indiana Jones had to overcome his fear of snakes . . . Over and over, audiences love to see this struggle in their stories.

And yet . . . if you're a scientist, you kind of want people to trust you by having them think your work is flawless. And yet . . . there are these things called error bars and error measurements and confidence intervals and all kinds of other signals that suggest the scientist is not a flawless character after all.

I know the science world worries a great deal about public image and keeping the public's trust. It's a difficult line to walk, and I'm not about to advocate that scientists eagerly share all their personal shortcomings with audiences. But what is important here in a practical sense is the power of the flawed protagonist concept in communication dynamics and the ways in which it can be used constructively. And scientists actually do use it quite often, whether they realize it or not.

The Powerfully Flawed Presentation

I have seen it many, many times in good science talks. The scientist is telling the story of an investigation and says something like "The mistake we made was rushing things. Every time we showed up to take samples, we felt we needed to rush them back to the laboratory. But finally one day we ran out of gas at our field site, and while we waited to be rescued we decided to collect a second set of samples an hour later, which was when we finally discovered . . ."

That's basically the exact story of the flawed protagonist. The flaw wasn't anything that would make you question the ethics or ability of the investigator. In this particular case, it was simply the natural human tendency to be in a hurry. But by telling it in a self-deprecating mode of "look at how foolish we were," the scientist can draw the audience in. They can relate, having made similar errors themselves.

In fact, one of the absolute greatest stories in the entire history of science is the discovery of penicillin. Its discovery was basically the serendipitous consequence of flawed work. In 1928 British biologist Alexander Fleming accidentally left open a petri dish of *Staphylococcus* bacteria, which became contaminated with blue-green mold of the genus *Penicillium*. He noticed that the growth of bacteria was inhibited around the mold. From this region he eventually extracted the first antibacterial compounds.

Despite the discovery, Fleming was a famously poor communicator and failed to convince anyone of the potential importance of the compounds. He published an obscure paper, and the knowledge languished for over a decade until picked up by the military and finally put to use in World War II. As a result, the story is usually cited as an example of the tragic consequences of poor communication.

But the element that is less appreciated is the storytelling power of the flawed protagonist—the scientist who commits an accidental error that eventually gives rise to a heroic discovery. If you made a bad decision in your research that eventually was corrected by a moment of realization, there may be more good than bad to the experience. Instead of feeling embarrassed about your mistake, ask yourself whether there might actually be some communications gold in the telling of that part of your story. So long as your intentions were honest, audiences will give you a lot of leeway with the basic idea that "everyone is human." Even scientists.

Developing Deep Narrative Intuition

The real goal is to develop a deep narrative intuition. You want to be able to just sense the problems with narrative. With intuition you can listen to someone talking and have a bell go off in your head, prompting you to say, "Wait, go back to that bit about the mistake you made in rushing too much. Tell me more."

This is what story development is about—looking at the facts and realizing that some elements are more interesting to an audience than others. Some have more dramatic content than others. A whole bunch of statements about how the work was conducted will eventually get numbing. *But* . . . (there's that word) . . . if you suddenly say, "But then one day we did things differently . . . ," that's when you're at last drawing on the power of narrative.

These things are not always obvious. That's the importance of developing intuition. This next one shows what you can miss out on if you haven't developed the intuition to spot the material with the most dramatic potential.

3. TAKING STOCK

A communications person at a major scientific institution told me the story of a group of their scientists who called in with a discovery. They had encountered a glacier that had recently melted. No one had ever reported it—it was news.

He said the people in their media office knew this was a potentially controversial finding, given the politics around climate change, but they decided to put out a press release. He went on to talk about the reception the press release received in the media, but I stopped him.

"Wait," I said. "Go back to the decision to put out a press release. Was that an easy decision for your group?"

Of course it wasn't. He told me about how their media office staff had split over it—half for issuing the press release, half for keeping quiet because of the inevitable politics of climate-related issues. There were terse and escalating discussions.

I said, "Okay, great, what else?" He said that night he and his wife also got into a spat as she sided with the "no press release" group—the opposite of his stance. And they were still arguing the next morning as he left for work. Lots more details followed, all of which were fun and fascinating, and the makings of good storytelling.

The key point here is that as he told me about all this material, it became clear that he had skipped past what the Logline Maker describes as the "taking stock" moment of the story. He was focused on the factual information of the story—how large the glacier was, what it's melting signified, what it meant for the future. All the things he felt were important. But when it comes to the mass audience, they are more drawn in by emotional content, and that's what the taking stock moment provides. So long as it is accurate, it is a valid part of the story and communications gold.

This is the point you want to get to—where you can listen to someone's story and have the intuition to hear these elements. Just as not all stories are created equal, the same is true of the parts of the story—some parts are far more narratively powerful than others. You just need to be able to spot them.

4. THE DARKEST HOUR

If you're using the Story Cycle template and you want to tell the story of a personal journey and have it reach deep inside of people, I've found sometimes the best starting point is to focus on element 6, the stage of the hero's crisis.

The "darkest hour" is often the most emotionally powerful moment of a story—that point where it looks like the protagonist is going to fail in his or her effort. A research institution recruited me to help tell the story of their nearly 30-year history of genuinely bold innovation. The start and end of the story were clear and obvious. It began with a dream of being a place for innovation, which is always a tough challenge. Today, the laboratory is hugely successful and admired. But the interesting question is, after the pursuit of the dream began, was there ever a time when it looked like the entire dream would fail?

In this case the answer definitely was "yes." There was a stretch when nothing seemed to be paying off and the pattern of failure might have resulted in the end of funding. But then . . . (there you go, the story begins).

There almost always is some period when things seemed to be unraveling and the journey seems destined for disaster. If you want people to really appreciate how amazing things are today, take advantage of the dramatic strength that comes from showing them how close it all came to failing. It's about the contrast and distance between the highs and lows.

Actually, all great storytelling is about the highs and lows. My first and most wonderful story instructor, screenwriter Christopher Keane, had a simple demonstration for this. He talked about telling biographical stories. He drew on the board a line that went way up then way down repeatedly. He said, "This is a graph of the person's life—lots of highs and lows." Then he erased all the middle parts of the line and said, "Now this is the material you want to use for your story. Spare us all the middle parts where he was in transition—just tell us about the highest and lowest points."

Furthermore, the more detail you can pull out of this stage ("the power of storytelling rests in the specifics"), the more powerful it will be. This is a lot of what constitutes good storytelling—the ability to sense when to speed things up and skip through the middle parts (a lot of informational details that will only interest the heavily quantitative types) versus knowing when to slow things way down and play out the drama and emotion that will pull in everyone—it's the baseball player swinging past two balls in slow motion, then hitting the grand slam. It's about draa-mah.

9

Results: The Narrative Spectrum

Now it's time to show how these templates work in the real world. In this chapter I use the Narrative Spectrum to evaluate the narrative structures of the abstracts of five published papers. In chapter 10 I present three cases from the past year of scientists using these tools to improve the narrative structure of their presentations. Finally, my colleague Stephanie Yin analyzes James Watson's classic account of scientific research, *The Double Helix*, using the Logline Maker Template. (Second disclaimer: As I said earlier, I am aware of the bad politics associated with James Watson—I use his work in this book only for purposes of analyzing narrative structure, not as any endorsement of his personal and professional life).

Analyzing Narrative Structure

We know that most scientists have never heard of the acronym IMRAD. Similarly, they probably don't give much thought to structuring the narrative of their research paper abstracts.

There are some journals, however, particularly in the biomedical world, that do guide the writers in creating a "structured abstract." In 2011 Anna Ripple lead a team of information specialists in analyzing the long-term patterns in this practice. They found a steady rise in the requirement of structured abstracts, from 2.5 percent in 1992 to 20.3 percent by 2005. As shown in figure 1, it took 50 years to adopt the IMRAD template in the biomedical world. It looks like a similar pattern is emerging for structured abstracts. The graph in their paper for the spread of this narrative feature, if extrapolated, shows that adoption is on schedule to hit 100 percent a little after the year 2050. Science changes slowly.

Ripple and colleagues grouped together all the various structure elements that journals ask for into five overall categories: background, objective, method, results, and conclusion. Background is the "and" material. Objective is the "but." Methods and results are the "therefore," and eventually it comes together with the conclusion. The same basic Hegelian thesis, antithesis, synthesis structure.

Now, to demonstrate the power and applicability of the ABT, I will apply it as a tool for analyzing the narrative structure of content, using the abstracts of scientific papers as the raw material for analysis. In January 2014 I gave the keynote address at the annual meeting of the Society for Integrative and Comparative Biology (SICB). They gave me a free subscription to their journal, *Integrative and Comparative Biology*. When my first issue (vol. 54, no. 2) showed up in the mail, it seemed only logical to choose the first group of papers in it—the proceedings of a symposium—to evaluate using the Narrative Spectrum.

The symposium had the great title "Parasitic Manipulation of Host Phenotype, or How to Make a Zombie." I analyzed only the abstracts of the papers. Just to be clear—the papers are all well written and clear. The only thing I'm pushing for is to reach a little higher for the full power of narrative. I could just as well have analyzed the entire body of each paper—narrative structure is something you want to have all the way through. But the abstracts provided simple, short passages for analysis. Here I present the abstracts for five of the papers, then share the result of my Narrative Spectrum analysis for each, followed by a few comments to support my designation.

ABSTRACT 1

We examined sand crabs (*Lepidopa benedicti*) for endoparasites, and found the only parasite consistently infecting the studied population were small nematodes. Because many nematodes have complex life cycles involving multiple hosts, often strongly manipulating their hosts, we hypothesized that nematodes alter the behavior of their sand crab hosts. We predicted that more heavily infected crabs would spend more time above sand than less heavily infected crabs. Our data indicate infection by nematodes was not correlated with duration of time crabs spent above sand. We also suggest that organisms living in sandy beaches may benefit from relatively low parasite loads due to the low diversity of species in the habitat.

ANALYSIS OF ABSTRACT 1: DHY

Abstract 1 begins by jumping right to a "therefore" statement, saying, "We examined . . ." This might almost work narratively if the title of the paper posed the question being investigated, which would enable you to jump right from the problem to the solution, but it doesn't. The title is a statement of results: "Nematodes Infect, but Do Not Manipulate Digging by, Sand Crabs, *Lepidopa benedicti*." The result is that with or without the title, there is no setup, just an immediate narrative direction. Then the second half of the first sentence presents results by saying, "and found the only . . . ," which means we're now getting results but still have no clear context. The second sentence begins with "Because . . ." This is another consequence word, like "therefore." Halfway through this sentence is "we hypothesized that . . . ," then the next sentence begins with

"We predicted . . ." By this point the narrative threads have bent backward and forward several times. There's your DHY dynamic in action. People who are familiar with the subject matter of this paper might have little trouble following it, but outside of that small group, the narrative is unnecessarily complex, to the detriment of readers.

ABSTRACT 2

Recent research suggests that plant viruses, and other pathogens, frequently alter host–plant phenotypes in ways that facilitate transmission by arthropod vectors. However, many viruses infect multiple hosts, raising questions about whether these pathogens are capable of inducing transmission-facilitating phenotypes in phylogenetically divergent host plants and the extent to which evolutionary history with a given host or plant community influences such effects. To explore these issues, we worked with two newly acquired field isolates of cucumber mosaic virus (CMV)—a widespread multi-host plant pathogen transmitted in a non-persistent manner by aphids—and explored effects on the phenotypes of different host plants and on their subsequent interactions with aphid vectors. An isolate collected from cultivated squash fields (KVPG2-CMV) induced in the native squash host (*Cucurbita pepo*) a suite of effects on host–vector interactions suggested by previous work to be conducive to transmission (including reduced host–plant quality for aphids, rapid aphid dispersal from infected to healthy plants, and enhanced aphid attraction to the elevated emission of a volatile blend similar to that of healthy plants). A second isolate (P1-CMV) collected from cultivated pepper (*Capsicum annuum*) induced more neutral effects in its native host (largely exhibiting non-significant trends in the direction of effects seen for KVPG2-CMV in squash). When we attempted cross-host inoculations of these two CMV isolates (KVPG2-CMV in pepper and P1-CMV in squash), P1-CMV was only sporadically able to infect the novel host; KVPG2-CMV infected the novel pepper host with somewhat reduced success compared with its native host and reached virus titers significantly lower than those observed for either strain in its native host. Furthermore, KVPG2-CMV induced changes in the phenotype of the novel host, and consequently in host–vector interactions, dramatically different than those observed in the native host and apparently maladaptive with respect to virus transmission (e.g., host plant quality for aphids was significantly improved in this instance, and aphid dispersal was reduced). Taken together, these findings provide evidence of adaption by CMV to local hosts (including reduced infectivity and replication in novel versus native hosts) and further suggest that such adaptation may extend to effects on host–plant traits mediating interactions with aphid vectors. Thus, these results are consistent with the hypothesis that virus effects on host–vector interactions can be adaptive, and they suggest that multi-host pathogens may exhibit adaptation with respect to these and other effects on host phenotypes, perhaps especially in homogeneous monocultures.

ANALYSIS OF ABSTRACT 2: ABT

Here's proof of the power of the ABT structure. The first sentence of Abstract 2 lays out clear exposition. The second sentence gets right down to business, starting with the contradiction word "however," which could just as easily be "but." The essence of the sentence is "are these pathogens capable of . . . ?" The next sentence is the "therefore." It begins, "To explore these issues." You could drop in "therefore" at the start and it would feel just fine. Overall, the abstract provides a very concise opening that is easy to follow and sends you off in a clear direction. From there it details the work (perhaps a little more extensively than is needed for just the abstract). Near the end, with the sentence beginning "Furthermore . . . ," we can sense we're getting near the end of the reporting of results. The next to last sentence begins with "Taken together . . ."—clearly it's wrap-up time. And the final sentence starts with "Thus, these results are consistent with the hypothesis that . . ." There's no doubt that this is rather wordy abstract that could have benefitted from some trimming, but it at least has a very clear, simple and strong narrative structure.

ABSTRACT 3

Animals have a number of behavioral defenses against infection. For example, they typically avoid sick conspecifics, especially during mating. Most animals also alter their behavior after infection and thereby promote recovery (i.e., sickness behavior). For example, sick animals typically reduce the performance of energetically demanding behaviors, such as sexual behavior. Finally, some animals can increase their reproductive output when they face a life-threatening immune challenge (i.e., terminal reproductive investment). All of these behavioral responses probably rely on immune/neural communication signals for their initiation. Unfortunately, this communication channel is prone to manipulation by parasites. In the case of sexually transmitted infections (STIs), these parasites/pathogens must subvert some of these behavioral defenses for successful transmission. There is evidence that STIs suppress systemic signals of immune activation (e.g., pro-inflammatory cytokines). This manipulation is probably important for the suppression of sickness behavior and other behavioral defenses, as well as for the prevention of attack by the host's immune system. For example, the cricket, *Gryllus texensis*, is infected with an STI, the iridovirus IIV-6/CrIV. The virus attacks the immune system, which suffers a dramatic decline in its ability to make proteins important for immune function. This attack also hampers the ability of the immune system to activate sickness behavior. Infected crickets cannot express sickness behavior, even when challenged with heat-killed bacteria. Understanding how STIs suppress sickness behavior in humans and other animals will significantly advance the field of psychoneuroimmunology and could also provide practical benefits.

ANALYSIS OF ABSTRACT 3: AAA

Abstract 3 is largely an "and, and, and" presentation. It's a review paper, but the abstract is one long chain of statements capped off with a final sentence that says all this research will "significantly advance the field of psychoneuroimmunology and could also provide practical benefits." Up until then there is simply no narrative thread, just an itemization of behavioral defenses against infection. Flash back to the section where I said there's nothing wrong with an AAA structure so long as it is accurate. But it does represent a missed opportunity for using narrative to raise it to a higher, more concise and compelling plane.

ABSTRACT 4

For trophically transmitted parasites that manipulate the phenotype of their hosts, whether the parasites do or do not experience resource competition depends on such factors as the size of the parasites relative to their hosts, the intensity of infection, the extent to which parasites share the cost of defending against the host's immune system or manipulating their host, and the extent to which parasites share transmission goals. Despite theoretical expectations for situations in which either no, or positive, or negative density-dependence should be observed, most studies document only negative density-dependence for trophically transmitted parasites. However, this trend may be an artifact of most studies having focused on systems in which parasites are large relative to their hosts. Yet, systems are common where parasites are small relative to their hosts, and these trophically transmitted parasites may be less likely to experience resource limitation. We looked for signs of density-dependence in *Euhaplorchis californiensis* (EUHA) and *Renicola buchanani* (RENB), two manipulative trematode parasites infecting wild-caught California killifish (*Fundulus parvipinnis*). These parasites are small relative to killifish (suggesting resources are not limiting), and are associated with changes in killifish behavior that are dependent on parasite-intensity and that increase predation rates by the parasites' shared final host (indicating the possibility for cost sharing). We did not observe negative density-dependence in either species, indicating that resources are not limiting. In fact, observed patterns indicate possible mild positive density-dependence for EUHA. Although experimental confirmation is required, our findings suggest that some behavior-manipulating parasites suffer no reduction in size, and may even benefit when "crowded" by conspecifics.

ANALYSIS OF ABSTRACT 4: DHY

Abstract 4 ends up being iconic for the Despite, However, Yet Template. In fact, I drew the template's three-letter abbreviation from this particular abstract. The first sentence, for starters, is 67 words—that's almost a whole abstract in one sentence! And it is not a simple one—it's a conditional statement built around "whether the parasites do or do not," which already has us going two directions. The second sentence begins with Despite. The third sentence begins with However. And the forth sentence . . . yep, sure enough, starts with Yet. This is what I'm talking about when I refer to communications being "overly narrative." The reader has been pointed off in four different directions in four sentences. I'm sure it's all accurate, so it's not like it's a disaster. There are just simpler ways to present this material. Simpler is not easy, but it starts by getting to know the ABT structure intimately.

ABSTRACT 5

Parasites that adaptively manipulate the behavior of their host are among the most exciting adaptations that we can find in nature. The behavior of the host can become an extended phenotype of the parasites within animals such that the success and failure of the parasite's genome rely on precise change of the host's behavior. Evolutionary biology was born from the close attention of naturalists such as Wallace and Darwin to phenotypic variation in seeking to understand the origins of new species. In this essay, I argue that we also need to think about the origins of parasite-extended phenotypes. This is a more difficult task than understanding the evolution of textbook examples of novelty such as the eyes of vertebrates or the hooves of horses. However, new tools such as phylogenomics provide an important opportunity to make significant progress in understanding the extended phenotypes of parasites. Knowing the origins of parasite-extended phenotypes is important as a goal all by itself. But the knowledge gained will also help us understand why complex manipulation is so rare and to identify the evolutionary tipping points driving its appearance.

ANALYSIS OF ABSTRACT 5: ABT (THEY SAY, I SAY)

Abstract 5 is an argument—as if the author had read and taken to heart Graff and Birkenstein's *They Say, I Say*. It opens with two clear and straightforward sentences, laying out the "they say" side of the issue. The third sentence does not start with a contradiction word, but if you drop one in you can feel it would work. Try it. "But evolutionary biology was born from the close attention of naturalists . . ." You could then add "Therefore" to the next sentence, having, "Therefore in this essay, I argue that . . ." So the "but" and "therefore" were not present, but the structure was, providing the tripartite form, clean and simple. Georg Hegel would be happy.

I did this Narrative Spectrum analysis for all 13 papers from the symposium. Here's how they fell out: ABT: 6; DHY: 6; AAA: 1.

Six of the abstracts hit on the ABT structure or came close to it, establishing a single source of tension or conflict, then heading off in that direction. Those abstracts don't suffer from being confusing or boring—they are delightfully understandable.

Six of the abstracts fell on the overly narrative end of the spectrum. They offered up more than one contradiction word, had a contradiction word as the opening word of the abstract, from the very start had multiple contradiction words in a row, or had the ABT elements in the wrong order, all of which cause confusion. I have to say that for many of the DHY abstracts, I had to read and reread and then reread again, trying to figure out what was being said. But that wasn't the case for the ABT-structured abstracts.

Only one abstract was an AAA—just presenting a series of statements with a single wrap-up sentence at the end. I have a feeling this ratio of the three types may be fairly reflective of the state of scientific papers generally. Half of veteran scientists instinctively land on or near the ABT structure, but then about half of them overthink things, resulting in multiple narrative threads and jumps in logic.

In a talk I gave to the US Department of Agriculture, I asked one of the scientists to randomly select a volume of one of the journals he uses the most. He gave me volume 35, issue 3 of *Systematic Botany*. I handed it off to my Story Circles co-producer Jayde Lovell, who quickly set to work seeing if all of the abstracts of the 19 papers in that issue had perfect ABT form. Her conclusion: not even close.

Using the 1 to 10 scale, with a 10 being perfect ABT structure, she ended up with an average of 3.8. There were two 8's and a 9, so it's not like they were all bad. But there were also three 1's. There is of course a significant subjective element to this assessment— someone else could easily have gotten an average of perhaps 6—but

no one would begin to suggest they all had solid narrative structure. Clearly there is plenty of work to be done.

One more anecdote to add to this: A friend read this manuscript
then asked her graduate students to bring to their next lab meeting two abstracts, "one good, one not good." That was all she gave
them for instruction. The first-year students chose abstracts based
on their content—basically, "This one is good because it's about an
interesting subject; this one is bad because the subject is boring."
But the "good" abstracts selected by the older students were mostly
ABTs, while their "bad" ones were just a narrative mess. She was
impressed. Even with no specific knowledge of the ABT, the older
students were drawn in that direction. But young scientists can do
better than just having a gut feeling about what works. My hope is
that, with the terminology and specifics of this book, they can build
solid narrative intuition and then learn to articulate that intuition
effectively. That's what gives you the real power of narrative.

10

Results: Four Case Studies

So do the narrative tools work? Good question. I had my concerns, but the answer is yes. Here are four illustrative case studies. The first two involve groups of scientists I have worked with as they put these tools into action. The third describes the success one of my former workshop participants enjoyed after implementing what she had learned. And finally, my colleague Stephanie Yin analyzes James Watson's classic account of scientific research, *The Double Helix*, using the Logline Maker Template.

The challenge is that every case is different. It's not like you just go to work with the WSP elements and bingo, you're done. In some cases only one or two of the elements are useful. But the fact is, the more you work with them, the more they transition from memorized cerebral elements to a more visceral and intuitive sense that you have for why something isn't working and how to fix it. And of course, in the end, it is all about problems and solutions.

Case Study 1: The Sea Level Rise Panel

I start with how we applied the tools to the sea level rise panel. After the two scientists and I had quickly dispensed with our minor interpersonal disconnect, we agreed on three core words that described the material, and we replaced the initial title "Responding to Sea Level Rise" with the more interesting "Sea Level Rise: New, Certain and Everywhere." It's a much more powerful, specific and memorable title—even reminiscent of Tom Friedman's bestselling book *Hot, Flat and Crowded* or Jared Diamond's *Guns, Germs and Steel*.

Once we had these three words, we created ABTs for each. In our first conference call we came up with the following:

NEW: For 8,000 years sea level has been stable AND civilizations have been built right to the edge of the ocean, BUT for the past 150 years sea level has been rising rapidly, THEREFORE it is now time to come up with a new management plan for coastal areas.

CERTAIN: Sea level rise is the result of human activity (atmospheric alterations) AND we do need to work on curbing greenhouse gas emissions to ultimately stop the source of the problems, BUT for now the train has left the station, meaning despite the impressions some people have given that we can still stop sea level rise it simply isn't true—we are certain some sea level rise is going to happen no matter what we do, THEREFORE while we continue to work on mitigation we must also set to work on adaptation.

EVERYWHERE: Sea level rise is having major impacts in distant locations like Micronesia AND the Mediterranean, BUT it's not only happening in those remote locations—it's happening all around the planet and in some places as much as 100 miles inland, THEREFORE we must get the public to realize this isn't someone else's problems—it's going to impact everyone and everywhere eventually.

Once we had these basic structures for the three stories, we then moved to the next level of detail, again creating ABTs. You could call them "nested ABTs."

For example, the start of the EVERYWHERE ABT makes reference to a story from Micronesia. Here's an ABT for that story, "Sea Level Rise has been happening throughout Micronesia AND breadfruit is a traditionally important crop, BUT now the water table is rising and damaging the breadfruit, THEREFORE growers are being forced to move the crops to a higher elevation."

By the time we were done we had a whole series of these nested ABTs. For the opening of the entire presentation, I spotted two great story pieces from one of the scientist's previous talks that I cobbled together into a single vignette that would set the tone of the event.

In the first story, the scientist related that, after Hurricane Katrina, former US Senator Mary Landrieu told the residents of her state she was going to visit the Netherlands, the country where they have figured out how to live safely with the ocean despite much of the country being below sea level. In another part of his presentation, he quoted a 2012 speech in which the Dutch ambassador said, "We have finally realized we can't always fight the ocean."

Joined together the two pieces made a great ABT. It was basically, "The senator said we've taken a beating from the ocean AND she said she was going to the Netherlands, where they have figured out how to fight the ocean, BUT the Dutch ambassador said we've come to accept we can't always fight the ocean, THEREFORE we are gathered here today to address this predicament of a rising ocean." It was the perfect vignette to set up the panel's theme of "You can't always fight the ocean." (Cue Dobzhansky—"Nothing in the issue of sea level rise makes sense except in the light of not being able to fight the ocean"—and there you have your message).

We put on the event and it came off tremendously. The three of us took turns telling the stories to the 1,000 people filling the ballroom. When it ended, the organizer of the meeting, Steve Weisberg of the Southern California Coastal Water Research Project, said he had organized or attended dozens of these panels in his career, but this one was the best by far and felt like it was at a higher level than the others (thanks to the power of narrative).

There were raves all around. Four months later, at another meeting, five people came up to me and thanked me for that panel session, saying they had never seen anything like it. Best of all, a letter I wrote about it was published in *Science* a month after the meeting. But let me tell you about the most important lessons.

YOU GET WHAT YOU PAY FOR

The night before the talk I went to dinner with the two scientists. They were both a little apprehensive, uncertain whether we would flop or fly. They assured me that regardless of what happened they were glad they had taken the chance to try something new. But more importantly, they both commented on how they had never . . . ever . . . invested that much time and energy into a presentation.

Over the six weeks before the meeting we had conducted four conference calls, numerous one-on-one calls, and traded countless emails. It was indeed a huge amount of work, which prompts the question, How important is it for you not just to be understood with a simple AAA presentation but to actually engross, entertain, provoke and engage an audience from start to finish with the power of narrative using the ABT structure?

It's a serious question, and in the past the answer unfortunately has been, "We don't care enough to feel it's worth the time and energy." But increasingly the answer these days is indeed, "Yes, let's do it," as scientists discover the power and importance of narrative. You just have to know it doesn't come quickly and easily. You basically get what you pay for.

EVEN THE BEST SPEAKERS NEED NARRATIVE HELP

One more point—needing help with narrative structure is neither unusual nor belittling. The two scientists from the sea level rise panel, though highly respected experts in their field, needed help with it. Everyone needs help with it at some point (you should have seen the start of this book before Jerry Graff gave me guidance). Everyone. This includes Steven Spielberg. It especially includes Ron Howard and Brian Grazer, the director/producer team behind the movie that inspired the title of this book, *Apollo 13*. I attended an ocean conservation banquet in Hollywood where they were honored for their efforts to help save the oceans. In their joint speech they talked about how together they did their first scuba dive in Hawaii, but it was the stumbling-est, bumbling-est, dullest, most ambling "And, And, And" presentation imaginable. They were bo-ho-horing. And I just sat there in shock thinking, "THESE are the great storytellers of Hollywood?" Everyone needs help with narrative structure.

In the case of the sea level rise panel, neither of the two scientists needed any help whatsoever with speaking and presentation skills. Both are experienced, charismatic, accomplished speakers. One of them has a great knack for humor; the other, a dramatic speaking presence that enabled him to close the presentation with a voice of authority that drove home the issue for the audience.

I just sat there throughout the event kind of marveling that neither of them needed me to "coach" them on how to make eye contact with the audience and have good posture. None of that stuff— they had it all down. The *only* thing they needed help with is what everyone needs help with—working and reworking the information to find the optimal structure. It's a giant exercise in puzzle solving that can be frustrating at the start but hugely rewarding by the end, if you put in the time to solve it properly.

A FOURTH VOICE

There was one more very cool element to the sea level rise presentation: we used the ABT to provide a fourth voice in the presentation— the voice of the audience. A month before the meeting I had the organizers send out an invitation to attendees: "Send us your ABTs on sea level rise." From around the country, scientists followed our enclosed instructions and sent us one-sentence ABTs that told specific stories from their own worlds.

The power of the ABT is what made this exercise effective. Had we said, "Send us one sentence," we would have gotten all kinds of unstructured, rambling material. Think back to my story about putting up the painting and asking people to describe it. With no narrative guidance, they rambled for almost 30 seconds. But with the ABT they produced a clear, concise sentence in less than half that time.

People need guidance and structure. Not a lot—just the little bit that the ABT provides. And that's why our request produced immediately useable results. At the end of each of the three overall stories, we presented three of the audience-submitted ABTs. This fourth voice added perspective to the presentation and also provided more specifics about the issue.

Case Study 2: The AAAS-Lemelson Invention Ambassadors

As I've said from the start, most of these tools I've developed are new, so I'm still learning how effective they are. I put them to the test for a project in 2014 and, somewhat to my surprise, they worked exceptionally well. (Hey, I'm a former scientist. I remain skeptical about everything.)

That summer a group from the American Association for the Advancement of Science (AAAS)—egged on by my buddy Shirley Malcom, head of education for AAAS—asked me to assist them with a new project they were developing with the support of the Lemelson Foundation called Invention Ambassadors. They had decided that each year they would select six scientist-inventors to work with as a team to help promote and foster the importance and role of invention in scientific research.

Their plan was to bring the team to AAAS Headquarters in Washington DC for three days. On the first day I would listen to their 12-minute presentations and give suggestions and recommendations on changes. The second day they would give the revised presentations to an audience of 200 administrators, program officers, venture capitalists, politicians and other interested DC folks. On the third day we would watch the videos of their talks and discuss.

Okay, now, having just heard about my sea level rise experience, can you guess what was wrong with that plan? Remember the meltdown with the two scientists? Remember my realization of how personal these presentations can be? Remember the six weeks we spent disassembling then reassembling the presentations?

Now think about the idea of notes on a Monday followed by revised talks the next day. It's a recipe for one of two things—either an emotional Armageddon or just not much revision. It also ran a huge risk of the speakers just blindly taking my advice—revising the structure of their talks then presenting something that they don't really know why they changed, other than "You said to change it." When that happens and it flops, you know where the finger of blame is going to be pointed—at yours truly. These things take time. A lot of time.

So we, the organizers, had a conference call two weeks before the event. I sounded the alarms then insisted on having the inventors' contact information. By the next day I had begun a series of individual phone calls with the six participants, initiating what would be a laborious process of working with them to structure their presentations. But it was also a wonderfully collaborative experience.

Here's the first thing I discovered: my former screenwriting instructor Frank Daniel was absolutely right. Remember his comment about how every first draft starts with the structure "and then, and then, and then"? Sure enough, each person was set to arrive in DC with exactly that structure—basically, "I was educated here and here, and then I did a postdoc here, and then I began working on this, and then I discovered this, and then I filed for a patent, and then we set up a company, and then . . ."

Nothing wrong with that material—all great for a first draft. And then . . . it was time to set to work.

The message at the core of *Don't Be Such a Scientist* was "arouse and fulfill," a mantra I first heard from USC communications professor Tom Hollihan in 1998. He said it's an age-old principle of mass communication that when it comes to reaching a broad audience, it's as simple as two things: "first you need to arouse the interest of the audience, then you need to fulfill their expectations."

So that was my first structuring principle for the Invention Ambassadors' presentations. Before each speaker dives into the details of what they discovered and what patents they received, let's have them begin with a story that will arouse the interest of the audience.

Next, think about the basic principle of "the power of storytelling rests in the specifics." What's the most specific story (and thus most powerful) that can be told about the invention process? Is it the story of five years or five weeks or five hours of research? No. It's the story of the one moment of invention—as small and finite as possible. It's that moment where everything comes together.

I have seen the power of this technique. My buddy Maggie Cary, who does communications training with the Mayo Clinic, told me about a great exercise she does with doctors. She has them "tell the story of the one moment in your career where you felt all your years of training coming together at once." My colleagues and I run this exercise with the doctors participating in our Society of Hospital Medicine workshops, and the results are amazing. The stories that come out are usually tales from the emergency room. A doctor tells about the one moment when he knew there wasn't time to research what was happening with a trauma victim—he simply had to draw on all his training, only to find himself solving problems he didn't know he was capable of solving.

So for each Invention Ambassador I began the conversation by asking them about the story of the one day, the one hour, the one moment when they felt their invention come together. The initial response was predictable. Each one of them said pretty much the same thing: "There was no single day or moment—it all happened slowly over the course of months and even years."

This is a pattern you often see when you start digging inside people's minds for stories—the tendency to default to generalities. I see it all the time when I'm filming documentaries. The process almost always begins with the subject answering questions with these sorts of generalities. It becomes the interviewer's challenge to dig deeper and get beyond the generalities. Which is what I began doing with each of these folks.

It was a little crazy after a while. Kind of like being a psychotherapist in search of "recovered memories." My experience with Steve Sasson is a prime example. He invented digital photography. That's right—he was working at Kodak in 1975 and was the first person to figure out how to use electricity to capture images. I began quizzing him in search of a "moment" of discovery. He did the standard thing of saying it all happened slowly. But then I began pushing (yes, I know, I'm obnoxious that way).

Slowly, slowly, bit by bit, like he was in the therapist's office, he began opening up, saying, "Oh, wait a second, now that I think of it . . ." He began homing in—he remembered it was 1975—it was that fall—it was actually in December—actually . . . wait a second . . . he pulled a journal down from his bookshelf, "Yeah, here it is—it was actually December 12, 1975—that was the day my assistant Jim and I finally had a complete prototype to try out."

He told "the story" of how he and Jim rolled the big cart of equipment down the hall to Joy, the receptionist. They asked her to stand against the wall, then they flipped a switch, rolled the cart back to the laboratory, hooked it up to a monitor and there, lo and behold, was an image. It was a blurry mess and they had made a minor

programming error—all the pixels were reversed so it was a negative image—but they could see something that to them resembled
Joy. It was indeed *the* moment I was seeking.

And then he said the laboratory door opened up and in walked
Joy, who took one look at the blur on the screen and said, unim-
pressed, "This is gonna need a LOT of work!" and left.

There you have it. A great, fun and perfect story with which to
draw in everyone in the auditorium, lighting the fire of their interest and opening the door to the "fulfill" side of the equation, where
they will eagerly embrace whatever information you provide (rather
than shying away because of information overload).

And look at where the power of the story comes from—not the
informational details of how the first prototype digital camera was
constructed but rather the emotional content of seeing the first
image, then the humor of the nonscientist criticizing what in the
eyes of the scientists was an object of beauty.

So uncovering each Invention Ambassador's "moment" of invention was the first step in helping them develop compelling presentations. But then there was a second step that was even more
powerful, as much for me as for them.

COMING FULL CIRCLE TO THE ABOMINABLE
ACTING TEACHER

If you look at the opening of *Don't Be Such a Scientist*, you'll see a full
paragraph of profanity, screamed by my monstrous acting teacher
on the first night of her class. I realized, over the years, that she was
the best, most effective teacher I've ever been exposed to (though I
can't give you a "closed ending" to this story as I haven't spoken to
her since 1996, when I finished the class). This Invention Ambassadors experience ended up being a moment of "coming full circle"
with her teachings.

After lengthy one-on-one conversations with me during the two
weeks before our DC visit, the Invention Ambassadors showed up

on Monday afternoon at AAAS for our first formal session. Instead of being my first exposure to their ideas for presentations, it was my opportunity to hear what they had created from the basic outlines we had already devised together by combining their raw material with my guidance in the narrative-shaping process.

We ran through each story. I gave them detailed notes to work on overnight, then the next morning we did a full rehearsal of their presentations before lunch in preparation for the formal presentations at 3:00. When they finished the rehearsals I ran through a final batch of notes with them. But this time I prefaced everything with the warning and apology that it was probably too late for them to incorporate much of what I had to say, and more importantly, that they shouldn't make any changes they themselves didn't feel like they wanted. It was essential that everything presented come from inside of them.

But here's the big thing that happened. I myself got hit by a "moment." I found myself having the same basic note for each speaker, and it took me back to my acting classes of nearly 20 years earlier. The note was "Enough with the telling us about what happened to you—we want to know what happened inside of you." It's kind of the same note as "Enough information, we want the emotion."

I began pushing them on sharing with the audience the emotional experience of these key moments of discovery. For example, when you finally tell us about seeing that first cloud of pixels on the screen, take a moment and tell us what they call in acting "your inner monologue"—which means basically tell us what was going on inside of you. Tell us, for example, about how your parents had always hoped you'd achieve some major accomplishment—about how "this was the moment I finally lived up to their hopes and dreams."

After listening to Vinod Veedu (who invented and patented the idea of "nanobrushes") tell all the "and then, and then" details of his career, I began by having him create a starting point for his journey by telling about what went on inside of him while growing up in India. He said all his friends became IT guys working in cubicles, on the phone all day, every day. He viewed that life as boring and made a vow to himself that he would never end up in such a rut.

He eventually moved to the United States to earn his PhD, but no sooner did he graduate than he found himself in a deadly boring job describing the surfaces of nanofibers. *But then one day* (the story begins!) he spotted something—a microscopic structure that looked like a brush. He decided to look at it under the scanning electron microscope. His colleagues predicted he would be wasting his time, but he sensed something. He looked into the microscope, adjusted the focus, zoomed in on the structure, and there he saw his own object of beauty—an infinitesimally small brush-shaped structure, which he named a "nanobrush."

At this point I gave him the same note I had given the others about internal monologue. I asked him to stop in his presentation, even turn to the audience, and describe what this all meant to him. Take us back to your childhood—to your fears of a boring life—to how long your struggle had been—and then tell us exactly what you felt at that moment.

Insights like these are the "communications gold" the audience is really seeking. And if you can give it to them authentically, they will make a deal with you—in return they will listen closely over the next few minutes as you delve into the science of how these nanobrushes are formed, why they are valuable, and what patents you have filed for them. And they will be interested because you emotionally aroused them. And even if they don't totally understand all the science, they will still do their best to listen.

EVERY LITTLE MOMENT HAS A MEANING ALL ITS OWN

But as I said, this was also a "moment" for me. It took me back to the fateful night in August of 1994 when I began my first acting class in Hollywood, the night when the paragraph of pure profanity spewed out of the acting teacher's mouth—her hateful rage against me on the very first night we met. It was an experience of confusion that would take me more than a decade to completely process.

That night my scene partner insulted me in the middle of an exercise. The acting teacher halted us, came running up, and shouted in my face (and I'm leaving out all the expletives), "How does that make you feel???"

I shrugged and with the totally analytical, nonemotional, indifferent voice of a scientist said, "I don't know, it's not that big of a deal."

That was when she lost it. She screamed, "*In this class you can be mad, you can be sad, you can be glad, but the one thing you CANNOT be is without emotions—nobody wants to listen to a person who has no emotions!!!*"

That was the moment, back then, that changed my life. And now this was the same moment again, two decades later, for me. A light switched on in my head. The ghost of that horrible woman was hovering in front of me as I gave (albeit more politely) this same note to the Invention Ambassadors—don't be an inhuman robot. Share something more than the cold, clinical facts with us. Give us something that will reach inside of us, tapping into our emotional side.

So guess what happened. Did they fail to listen to me? Did they argue and negate what I was saying? Did they reject my advice? No. They took all my notes and incorporated them into their talks.

Three hours later as they gave their formal presentations, I sat in the back row of the theater listening almost in shock and in all seriousness with tears in my eyes at this moment of revelation. Remember when I said back in part 2 that scientists don't listen? Remember the reviewer of *Don't Be Such a Scientist* who even said this in *Science* magazine? These scientists listened. What happened? I'll come back to this near the end of the book, when I talk about building the perfect scientist.

Case Study 3: A Foote Note

The ABT often brings instant gratification. Many people take one look and within minutes put it to use. At the end of a talk I gave at Princeton University, a graduate student started the Q&A by saying, "As you were speaking I worked out the ABTs for each of the chapters of my dissertation. I wish I had been I taught this three years ago— it would have been such a help."

So many successful uses. Here's another example from the many emails I get from people who have put the ABT to work for their communications efforts and reaped the expected benefits.

Liz Foote is the executive director of Project S.E.A. Link, working to preserve the ocean habitats of the Hawaiian Islands through the creation of Marine Protected Areas. In the spring of 2014 she wrote to me after giving a big talk at the Ocean Sciences Meeting in Honolulu.

She talked in detail about how she used the ABT in assembling her presentation. She said, "My presentation essentially consisted of a bunch of mini ABTs embedded throughout. And mostly through photos; I used very little text. Surprising to me, I was actually enjoying the process of developing the presentation, and dare I say *excited* to have the chance to give it, rather than going through the usual motions of PowerPoint drudgery."

She went on to say, "I also did something I usually don't manage to do sufficiently before presentations—I practiced it a lot. I felt that I'd raised the stakes for myself and wanted to have the content locked down so I could focus on engaging the audience, showing some personality rather than "just trying to get through it." The reformatted structure using the ABT actually made it easier to remember the content, so I could convey it while sounding like a human and not a droning robot dependent on notes."

That last comment is really important—that the reformatted structure made it easier to remember. That's the circularity of the

ABT. Your brain is programmed to think, argue, reason and remember this way. Once you hit the ABT structure, it's not only easier for people to follow your argument, it's also easier for you to remember it.

She added, "I want to say thank you, for inspiring me to take something—a 15-minute presentation—from the standard "here's a bunch of stuff that we did and why," which I could have just phoned in on any given day, to an engaging presentation with a more compelling structure that obviously resonated with people."

The clincher for her was her "Sad Keanu" slide. As she told about how poorly designed the current signs are for a specific Marine Protected Area on Maui, she showed a slide with a photo of Keanu Reeves looking sadly down at the signs. (Why Keanu? Google "Sad Keanu" for the meme.) In closing her email to me, she said, "And as a bonus my Sad Keanu slide even got a laugh from enough people to be audible and bolstering, not embarrassingly awkward."

What more could you ask for? When you're killing it with your Sad Keanu slide, you're really reaching your full communications potential.

Case Study 4: James Watson and the Hero's Journey

The very best book of true science I have ever read is James Watson's *The Double Helix*. I read it long ago as an undergraduate yet still remember portions of it in vivid detail—especially the plot twists as Watson and Crick were competing against other laboratories to discover the structure of DNA. Remembering a story vividly many years later is almost always an indication that it had good narrative structure.

In fact, in *Don't Be Such a Scientist*, at the end of the chapter on storytelling, I retell the amazing story Neil deGrasse Tyson told in 2008 at a Hollywood event. It was about his first viewing of the movie *Titanic*. His story had perfect narrative structure—a tale of birth (he went to the movie and loved it), death (he saw they had the wrong constellations in the sky in the scene of the ship sinking), and rebirth (the director eventually fixed the movie thanks to him).

A year later I told the story at the start of a workshop. On the third and final day of the workshop I asked if anyone remembered the story. Every hand went up, and the one person I called on regurgitated it perfectly. All of which demonstrates the circularity of hitting perfect narrative structure—people take it in flawlessly, then remember it accurately. Once again, it is why this stuff is so powerful and important.

So it hit me—is there any chance, given how much I enjoyed Watson's book and how well I retained its contents (given the countless number of science books I've read for which I can hardly tell you the main point), that *The Double Helix* conforms to the Hero's Journey?

To explore this I had the wonderful and incredibly bright Stephanie Yin (whom I mentioned earlier—the recent Brown University graduate who worked with me for a year before heading to graduate school in journalism) read the book and see if it matched the Hero's Journey template. In a nutshell, it matched, big time.

Here's what she wrote, which I then posted on my blog:

"THE MOLECULAR BIOLOGIST'S JOURNEY: BREAKING DOWN *THE DOUBLE HELIX*,"
By Stephanie Yin

Reading *The Double Helix*, I was struck by the candid nature of Watson's writing. He became immediately familiar to me as a person, and this, in turn, made reading the book much more enjoyable—as if I were reading letters from a friend. Watson has all the trappings of *a flawed protagonist*: he is young, foolhardy, searching for fast shortcuts to fame and seduced by the world of the educated, European socialites around him. His flaws set him up to undergo the

Hero's Journey. Below is a summary of this journey, using the language of the Connection Storymaker Logline.

In an ordinary world, a flawed protagonist

In an ordinary world, James Watson is a young scientist at the University of Chicago, primarily interested in studying birds, impatient for fame and looking for career shortcuts (in particular, avoiding taking any advanced chemistry, physics or math courses).

Feeling unfulfilled by ornithology, he becomes curious about how genes work. He starts grad school at Indiana University, advised by microbiologist Salvador Luria. At this point, he is interested in studying DNA but still hoping to avoid learning any deep chemistry.

A catalytic event happens

Watson has his life upended when, in the spring of 1951, he goes to a conference in Naples and hears a talk on X-ray diffraction of DNA by Maurice Wilkins, a physicist and molecular biologist at King's College. Around the same time, Watson realizes that these conferences were as much a gateway into a fashionable social scene as they were an entry into academia. He writes, "An important truth was slowly entering my head: a scientist's life might be interesting socially as well as intellectually."

After taking stock, the hero commits to action

After taking stock, Watson becomes determined to learn chemistry and solve the structure of DNA. He decides to go to the University of Cambridge to learn X-ray crystallography. There, he meets and bonds with Francis Crick, who is also interested in DNA. Watson writes, "From my first day in the lab I knew I would not leave Cambridge for a long time. Departing would be idiocy, for I had immediately discovered the fun of talking to Francis Crick. Finding someone in Max [Perutz]'s lab who knew that DNA was more important than proteins was real luck. . . . Our lunch conversations quickly centered on how genes were put together."

Together, Watson and Crick commit to finding the structure of DNA using a combination of X-ray photography and model building, a method that had recently been used by the biochemist Linus Pauling to understand the structure of proteins. "Within a few days after my arrival, we knew what to do: imitate Linus Pauling and beat him at his own game," writes Watson. "Now, with me around the lab always wanting to talk about genes, Francis no longer kept his thoughts about DNA in a back recess of his brain. . . . No one should mind if, by spending only a few hours a week thinking about DNA, he helped me solve a smashingly important problem."

The stakes get raised

After a while, Watson and Crick think they have stumbled across a breakthrough. They believe DNA is a three-chain helix with phosphate groups held together by Mg^{2+} ions. However, when Maurice Wilkins and Rosalind Franklin (who were studying DNA at the same time) visit Cambridge at Watson and Crick's request, they quickly find holes in this three-chain theory. Their idea thoroughly shot down, Watson and Crick are discredited, and their superiors order them to stop spending their time on DNA. "By this time neither of us really wanted to look at our model. All its glamor vanished, and the crudely improvised phosphorus atoms gave no hint that they would ever neatly fit into something of value," writes Watson. "The decision was thus passed on to Max that Francis and I must give up DNA."

The hero must learn the lesson, to stop the antagonist and achieve the goal

In order to find the structure of DNA before his competitors (Maurice Wilkins, Rosalind Franklin, Linus Pauling), Watson must learn to take his time, cultivate a deeper learning of chemistry and math-

ematics and resist his temptations to take shortcuts or rush to conclusions. For a while, Watson and Crick do their DNA research on
the down-low while making progress on their primary research.
(Watson focused on the structure of tobacco mosaic virus.)

During this time, Watson devotes a great amount of time to
learning chemistry—combing through scholarly journals and sem-
inal books on the topic. "I used the dark and chilly days to learn
more theoretical chemistry or to leaf through journals, hoping that
possibly there existed a forgotten clue to DNA," he writes. "The
book I poked open the most was Francis' copy of 'The Nature of the
Chemical Bond.' Increasingly often, when Francis needed to look up
a crucial bond length, it would turn up on the quarter bench of lab
space that John [Kendrew] had given to me for experimental work."
Watson hones his X-ray photography skills, thinks about DNA late
into his evenings and continually checks with reference books and
colleagues to make sure his chemistry is correct.

By the time he and Crick believe again that they have cracked
DNA's structure (which, of course, this time they had), they are vig-
ilant about checking their assumptions and obtaining exact coordinates before spilling the news, having learned from their earlier
fiasco with Wilkins and Franklin. "Keeping King's in the dark made
sense until exact coordinates had been obtained for all the atoms.
It was all too easy to fudge a successful series of atomic contacts
so that, while each looked almost acceptable, the whole collection
was energetically impossible," writes Watson. "Thus the next several days were to be spent using a plumb line and a measuring stick
to obtain the relative positions of all atoms in a single nucleotide."

By the end of the book, Watson and Crick have successfully predicted the structure of DNA, and it seems Watson has matured both
as a scientist (in his deeper grasp on chemistry and math, as well
as in his patience and restraint) and as a person (who is perhaps
no longer as taken with instant fame and the charms of the social
elite).

He ends the book in Paris, on a trip with his sister. In the last sentences of *The Double Helix*, he writes, "Now I was alone, looking at the long-haired girls near St. Germain des Prés and knowing they were not for me. I was twenty-five and too old to be unusual." On that note, our hero turned the page toward a new journey.

And there you have it—*The Double Helix*, a case study in the power of the Hero's Journey (as well as an extremely efficient piece of work from Stephanie Yin in breaking it down).

Some might argue that the enduring popularity of Watson's book is due to the superlative nature of its content (meaning it is the story of what might be *the* most important discovery in the history of biology—a story so interesting it doesn't matter how it is told), but they would be revealing their lack of understanding of the power of Joseph Campbell's work. In the hands of someone less narratively adept it could easily have been yet another dull account following the AAA structure, as are so many tales of scientific research.

Watson is different—he has the narrative intuition that I am advocating all scientists develop. He brought to that book the same narrative power that Stephen Jay Gould brought to his 25 years of monthly essays in *Natural History* magazine. Such is the power of narrative intuition.

A Final Note on Today's Crazy Kids

One last note here before we wrap up the Antithesis and move on to the grand Synthesis. It's about kids. They tend to get this story stuff better than adults. They live their lives in the world of story. They have yet to have the tsunami and fire hose of information of today's society replace so much of their emotional content with information. They are less negating, more open to affirmation of the sort that begins a story with a series of "and" statements.

I saw this a couple of years ago when I did a brief workshop that the National Academy of Engineering sponsored for a dozen superstar sixth to ninth graders. They were selected as part of a national competition called the Disney Broadcom MASTERS program. I was asked to spend an hour helping them craft the presentations of their individual projects.

As part of my exercise with them, I put the Logline Maker Template onto a single page with blanks to fill in for the nine elements of their individual projects. I handed out the sheets of paper, then came back to the front of the group to explain it. But to my surprise, before I could begin speaking, some of them were already halfway through filling out the forms.

As they were first looking at the sheets, I had been hearing comments from them along the lines of "Oh, yeah, this thing." And that was the deal. When they saw the word "protagonist," they instantly knew this was something from the world of story and, in particular, something akin to the superhero tales that inundate their lives.

It was a rather stunning contrast to what I get with adults in my workshops. With the adults you hear lots of "Hunh?" "What is this?" "How does this relate to what I do?"—often tinged with apprehension and nervousness.

With those kids it was just "Oh, yeah, this thing." And that's the experience I have with kids generally. It's worth keeping this in mind as you're considering these elements of story. It was once all so easy for you when you were kids. What happened, and how do we get back to that?

It's almost a recapitulation thing—seeing our early cultural history in our early developmental stages. Almost. Maybe. Okay, on that note (before I open up a huge can of evolutionary biology worms), let's move on to the grand synthesis.

IV

SYNTHESIS

ABT—Always Be Trying . . . New Things

At dinner the night before the sea level rise panel, the two scientists and I chuckled and speculated on how badly our session might go. There had been no time for rehearsal, and, with everything locked into a Prezi file, there could be no last-minute adjustments. Nonetheless, we toasted to having at least taken on something different.

Most significant to me about the entire experience was that the success of our collaboration had been possible only because the scientists dropped their guard a bit and trusted me, despite my having been contaminated by living in Hollywood. They got beyond their fears. Their intuition told them I knew what I was doing. And so this is what I want to begin with—a synthesis not just of this book but of my 25-year journey from academia to Hollywood—by talking about the elusive human traits of fear and intuition.

I gave up tenure. In case you're not entirely clear what tenure means, it is the guarantee of a position as a professor at a particular institution, with all the benefits of health insurance and retirement, for the rest of your working life. It is the golden chalice that academics seek—the crowning achievement of most successful professors.

In the more than 20 years since I left my tenured professorship in the field of marine biology, by far the number one thing that people have asked me is, "Weren't you afraid to give up the security of tenure?" The closer they are to the academic world, the more they find it baffling, stunning, mindboggling, almost logic defying that I did such a thing. They ask, "Why would someone work so hard for something so difficult to achieve and which brings with it such a gift of lifetime security, only to give it back as soon as it is earned?"

I had a multitude of reasons. Some were personal, such as my divorce. Some were professional; I felt I had accomplished and experienced most of what I dreamed of doing as a marine biologist. But the single best explanation of what caused me to make the move is that I was guided by my intuition.

That may be the worst thing for many scientists to hear, but let me explain a bit further. I have always had fun at parties telling about how I "threw away tenure," as though it were an impulsive, crazy act of rebellion in which I ran out the doors of the Biology Building at the University of New Hampshire shouting iconic lines from movies and never went back.

As you might imagine, the truth was far from that. I actually approached my career transition with the systematic acumen, curiosity and thoroughness of a true scientist. In my second year as a professor I began making exploratory trips to Hollywood—flying out from Boston on Wednesday nights, staying in the Hyatt Hotel on Sunset Strip, doing meetings in Hollywood during the day, attending parties and dinners at night, learning the lay of the land, then flying back on Sunday to be ready for lecture Monday morning. Over the course of four years I probably made eight such trips.

At the same time I was making short films, winning awards at film festivals, taking film workshops in Boston and Rockport, Maine, writing book manuscripts and screenplays, and overall just hustling endlessly while still teaching and conducting marine biological research. By the fifth year of my professorship there was little mystery about where I was headed. I wanted to delve deeper into the mass communication of science and I knew where I needed to go. I applied to USC Cinema School and was accepted.

But there were still no guarantees. There were no data that could give me 100 percent certainty of succeeding with what I was about to do. At some point I had to gather together all that I had learned in my exploratory work and then rely on what my intuition told me—that it was worth taking the leap.

There were a few tough times, which I recounted in *Don't Be Such a Scientist*, but overall things went pretty much according to expectations. I think this ends up being perhaps the main reason why I am so firmly convinced of the power and importance of intuition. The world is not always entirely knowable for any given complex set of problems (just ask the climate scientists). At some point there has to be the ability to synthesize information at the higher level that our brains enable us to achieve.

Intuition is the only hope for overcoming the unknown in a way that doesn't involve fear. It becomes the one-word summary of my entire journey. Nothing in my journey makes sense except in the light of intuition.

So just to review, science now faces problems with scientific research (false positives and a bias against null results) and with science communication (delivering boring presentations at best, unintentionally fostering antiscience sentiment at worst). Underlying both is a lack of narrative intuition. Now it is time to get beyond the fear of Hollywood, beyond the fear of story, and improve the narrative intuition of the science world.

11

Science needs story . . .

It's Still a Narrative World

When it comes to narrative, not that much has changed in 4,000 years. It was a narrative world when they carved the story of Gilgamesh into stone; it's still a narrative world today. All day, every day, you are living in a narrative world. When you listen to your friend tell about her family's trip to Europe last summer, when you hear the news, when you watch television—all day long, narrative upon narrative. Yes, we're communicating more rapidly, but story still rules. Just ask any successful moviemaker today.

Proof of the enduring power of story can be seen in a 2014 study by Keith Quesenberry, a researcher at Johns Hopkins University who examined the content of Super Bowl commercials for the previous two years. He found that despite all the cute animals and sexy bodies, the most important factor accounting for the overall success of various commercials was still—you guessed it—the strength of the storytelling.

Here's a comparative example of the power of story. Two feature films in the United States have addressed the issue of global warming and managed to find large audiences. One was *An Inconvenient Truth*, a documentary film in which former presidential candidate

Al Gore gives a lecture on how we have altered our atmosphere and what the consequences may be. It is pretty much an "And, And, And" presentation.

The other movie, *The Day after Tomorrow*, is a fictional story of a world dealing with a climate drastically altered by human activities. Good science is nowhere to be found in that film, but there is good narrative structure. It has a solid and suspenseful ABT at its core. It's like the Hitchcock clips in Hasson's fMRI studies.

The AAA movie made $25 million at the box office. The ABT movie scored $186 million. People still like a good story.

Yes, the fiction movie was cockamamie and packed with bad science. At the sold-out screening I attended in Los Angeles the audience howled with laughter as Dennis Quaid delivered utterly silly dialogue as paleoclimatologist Jack Hall. Despite that, the movie was a huge success. This shows how powerful story structure continues to be as a force of successful communication, regardless of content.

It is still a narrative world, with story woven into virtually everything. So why fear it? This may be the most important question I pose in this entire book.

Storyphobia—The Irrational Fear of Story

I think the term *storyphobia* is new with me. A search on Google shows no signs of it. The term is needed. As I mentioned earlier, I have suffered its consequences (as when a *New Scientist* reviewer incorrectly accused me of advocating "bending the science to tell better stories").

In 2013 *Nature Methods* published a series of editorials about the role of "storytelling" in the writing of scientific papers. One of the editorials, "Against Storytelling of Scientific Results" by MIT neurobiologist Yarden Katz, shows clear evidence of storyphobia. Katz offers an impassioned plea to keep "storytelling" out of science. He says, "Great storytellers embellish and conceal information to evoke a response in their audience." But what does he mean when he says "storytellers"?

If you accept my premise that everything tracks back to the tripartite structure of the Hegelian dialectic, then it makes no sense for Katz to cast aspersions on the tellers of stories. He should have been more specific. He should have said, "In using the word 'storyteller,' I am referring to the tellers of untruths."

With this, I think we are finally getting to the core stumbling block for science when it comes to story. There is a lack of clarity on the meanings of the words *story*, *storytelling* and *narrative*. I have been mentioning them all along in this book, but any specific definitions I could have offered wouldn't have made sense without context. Now that you have that context, let's pin down the definitions.

Defining *Story, Storytelling* and *Narrative*

My Connection Storymaker workshop co-instructor Brian Palermo began jabbing this knife into my side after a few runs of our workshop. He began telling me that I needed to define what exactly is meant by the word *story*. At first, before I had come to a full understanding of what we were facing among our workshop participants, I argued that there is too much art involved in the whole concept of "story" to pinpoint it analytically with a formal definition. You might be largely in agreement with that defense. Perhaps you have gotten this far in your reading and still do not have a loud voice in your head shouting, "What exactly do these three words mean?" Everyone roughly understands what "story" means, right?

Actually, stop right there. Don't go one thought further. This is where I draw the line. No, they don't.

Yet that exact notion is commonly assumed, even in the scientific literature. Specifically, I'm talking about a 2014 paper by communications professor Michael Dahlstrom. He opens by saying, "Storytelling often has a bad reputation within science." His source for this statement is the Katz editorial. OK, true enough. That's part of the reason for this book. But then Dahlstrom goes on to be a shining example of the *actual* problem when he writes, "Most individuals have an inherent understanding of what it means to tell a story."

Again, no, they don't.

Houston, We Have Found the Problem

This is the single biggest message I have for you here at the end of my journey. The whole problem and challenge of story in science is that scientists often don't know what story is. Most people communicating in the AAA mode actually do think they are telling a story, but they aren't. At least not according to the definition I have arrived at, thanks to Brian's prodding.

I define "a narrative" or "a story" as "a series of events that happen along the way in the search for a solution to a problem." Think back to Campbell's circle diagram for a story—it was about problem-solution. And remember how I pointed to the parallel of the scientific method being an exercise in problem-solution.

This then means a "storyteller" is just someone who recounts the series of events that happened along the way in the search for a solution to a problem. Now we can start to see where the "problem" lies for the role of story in science. Someone stuck in AAA mode hasn't stated a clear problem and is not telling about a series of events on the way to solve a problem. To the contrary, they are just telling a bunch of information—just manufacturing perfect bricks in the brickyard with no idea of what those bricks are for.

That a paper published in a prestigious scientific journal (Dahlstrom's paper was in the *Proceedings of the National Academy of Sci-*

ences) starts off with the blanket assumption that everyone knows this stuff is a direct consequence of the core problem. It's similar to scientists having no idea what the IMRAD acronym stands for. Neither of these things is lethal to the doing of science. They are just *reflective* of a profession that has been resistant to using narrative as a tool for science communication and blind to where they are already doing just that (IMRAD). In fact, here's a little irony: if you look at the opening paragraph of the editorial by Katz, you'll see it has the standard ABT structure—the second sentence begins with "however."

There is currently a great deal of arm waving going on about story, narrative and storytelling in the science world, but the bottom line is that I fail to see anyone taking this sort of critical approach to these terms. The word *story* is being thrown around widely these days by people who use it to refer to just about anything anyone is saying. And I do mean just about anything. As in, "The professor started telling us the story of how molecules are assembled in the construction of G proteins." If all the professor did was list a series of facts, he wasn't telling you a story. There's your story problem.

Agon: The Cure for Storyphobia

Now is the time to rid science of storyphobia. Science is based on the gathering of knowledge through observation and experiment. It is meant to be logical and rational. There is no logical reason to fear the words *story, narrative* and *storyteller*.

Yes, there is indeed every reason in the world to fear the words *fraud, fabrication, deception, deceit* and *exaggeration* when it comes to science. By all means. And there are narratives/stories that can be filled with these qualities. But there are just as many others that are accurate, honest, true and reliable.

In the end, narratives are narratives—inert, constructed out of elements of logic. The ancient Greeks knew this. It was at the heart of their development of theater, which was built around the word *agon*.

Agon refers to a debate or a contest built around "a thing." It is basically the pursuit of a problem that has alternate, opposing solutions. The Greeks wrote their plays along these lines, and they derived two key words to describe the two sides—*protagonist* and *antagonist*. When the Greeks created the words, their entire concept of theater was not as an exercise between good and evil, but rather as two sides of, two approaches to, a thing, in the search for the truth. No moralistic values were assigned to either side.

That was Greek theater—an exercise in the search for the truth. It wasn't until centuries later, after the Renaissance, that the church resurrected theater in the form of what came to be known as morality plays, passion plays and miracle plays, where the terms *protagonist* and *antagonist* took on the moralistic values of good and evil.

So I say it's time to revert back to the Greeks. Isn't that what science is about—the search for the truth, independent of moralistic elements? Storyphobia in science is as misguided as the church's reworking of the Greek concept of theater. It is irrational and needs to be identified as unhealthy for science in general.

Addressing the challenges of science and story must begin with the acceptance that there is nothing intrinsically good or evil with the terms *story*, *storytelling* and *narrative*. Nothing. They are as value-free as $E = mc^2$.

So if we can agree that story is not something to be feared, that it underpins virtually everything and that the science world needs help with it, then let's define our problem as the need to bring greater understanding of narrative, of story, to the world of science, then go in search of the solution.

12

AND Hollywood can help . . .

I firmly believe Hollywood has the practical knowledge that can make a difference in dealing with these problems of narrative structure. But the cultural divide between Hollywood and academia is significant. From the first moment I visited Hollywood, I felt the discrimination against academics. While in film school I eventually ended up hiding the fact that I have a PhD. An advanced degree acts as a red flag that you are "caught up in your head," you are "overly cerebral," and you overthink things—none of which are good traits in the movie business.

Keep in mind these elements of divide I'm talking about are things you see when you live and work in Hollywood. If you are a scientist and come to a one-day event pulling together scientists and Hollywood people, you might think, wow, they are all so great, so friendly and interested in science. But it's different when you experience it over the long term. Very different.

Fortunately, science has been a central part of moviemaking from its beginnings in the late 1800s, when Eadweard Muybridge first rolled out his zoopraxiscope, one of the first devices for displaying motion pictures. Over the past century the importance of science in moviemaking has only increased. Today, the National

Academy of Sciences has formed a partnership with Hollywood in their Science and Entertainment Exchange program (a sort of matchmaking program that helps movie and television professionals find the scientific consultants they need), which I've taken part in since 2008. Most filmmakers are fond of science, and scientists are typically intrigued with filmmaking.

So you'd think that, given this historic and continuing connection, the culture gap between Hollywood and science wouldn't be that great. But it is. The dismissal of academia on the Hollywood side is mirrored in a disdain of Hollywood on the academic side, including the sciences. Which means that, before you continue reading, I need you to set aside any prejudices against Hollywood and open your mind to what I'm about to present. It has considerable conceptual power, but only if you can see the parallels and connections.

Back by Unpopular Demand: McKee's Triangle

I have talked about why positive results for scientific studies get more attention than null results. Now, in the spirit of "Dude, it's all the same story," I want to put this fundamental aspect of science into the bigger picture. I want to connect it with Hollywood by tapping into your preexisting knowledge and intuition about the box-office success or failure of movies. There are powerful similarities between the dynamics of how movies fare with the public and how scientific studies fare with both the public and the scientific community.

Your narrative intuition may not be at the level of a Hollywood screenwriter, but you definitely have at least some. Whether or not you have been a voracious reader of novels throughout your life, you have definitely watched a ton of movies. Everyone in American society has. It's the American way. You might think most of those movies were wasted time, but this is a chance to get something useful out of all those hours of viewing.

From *Schindler's List* all the way down the cultural ladder to the Transformers franchise, all box-office hits have one thing in common—their narrative structure. When a movie is released a predictable set of dynamics kicks in that is at least partly a function the movie's narrative structure. The structure will almost invariably predict the box-office success of the movie. At least at the level of blockbuster versus art house crowd sizes.

Now here's the parallel. When scientific studies are presented to the public, and even to the scientific community, this same set of dynamics kicks in. Really.

To convince you of this, I'm going to have to do something painful. I'm now slipping back into professor mode, where it's time for discipline as I tell my students I'm very disappointed with them.

In *Don't Be Such a Scientist*, I discussed something that I thought was both interesting and maybe somewhat important—McKee's Triangle. The concept comes from Hollywood screenwriting guru Robert McKee. It's detailed in his book *Story*, which is widely regarded as the bible of screenwriting in Hollywood.

In the five years since my book came out, here's how many people have commented to me on what I said about McKee's Triangle, either in person or in writing: zero.

Five years. No evidence that anyone even read it. The perfesser is not happy.

Meanwhile, I have developed an even deeper appreciation for McKee's Triangle. I see the concept everywhere these days and feel it is an enormously important tool for making sense of much of the world. So there's only one thing I can do at this point, which is to pull myself together and . . . SHOUT IT AT YOU THIS TIME!!!

Listen up, people, this thing is really, really profound and important. I feel certain that if you can "grok" its true meaning, it could change your life. (Don't know that word? Google it then read Robert Heinlein's classic book, as it, too, could change your life.)

Building Narrative Intuition for Science through Movies

So here we go again with McKee's Triangle (figure 13), the corners of which represent what McKee considers to be the three pure story forms. The lower right corner, *antiplot*, is the least important for our discussion. Purely antiplot stories strike out against all the constraints of structure and tradition. This is the place for revolutionaries who are fighting against the establishment just for the sake of the fight, not caring much about the outcome or how many people they connect with.

Included in this category are some of my favorite artsy movies that I watched in the quietude of New Hampshire (before I moved to Hollywood and my attention span shriveled up). Movies like *Meshes of the Afternoon* (the iconic abstract expressive short film we had to watch about 50 times in film school), *Stranger Than Paradise*, *Un Chien Andalou* (love the films of Buñuel), and some classic comedies like *Wayne's World* and *Monty Python and the Holy Grail*. All of them take traditional storytelling and stand it on its head. They are essentially nonnarrative.

The top of the triangle, *archplot* (pronounced "arc-plot"), is the most important story form for our purposes. McKee defines archplot stories as having the elements of "classical design." He says, "These principles are 'classical' in the truest sense: timeless and transcultural, fundamental to every earthly society, civilized and primitive, reaching back through millennia of oral storytelling into the shadows of time. When the epic *Gilgamesh* was carved in cuneiform on 12 clay tablets 4,000 years ago, converting story to the written word for the first time, the principles of classical design were already fully and beautifully in place." (Yay, Gilgamesh!)

Pretty sweeping statement. Archplot is the form that is age-old

ARCHPLOT

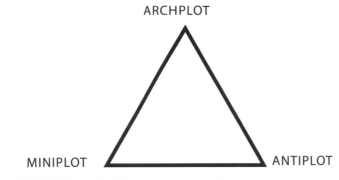

MINIPLOT  ANTIPLOT

Figure 13. McKee's Triangle. Back by unpopular demand.

and, not surprisingly, connects with the largest audiences. It's at the core of just about every movie that is popular with the masses. *Star Wars, Lord of the Rings, Gone with the Wind, Iron Man*—they all have an archplot structure.

McKee lists the major characteristics of archplot. Here are the five that are the most important and relevant to our discussion:

1. Linear Timeline—Events happen in sequence; they don't jump around.
2. Causality—Things happen for logical reasons, not randomly.
3. Single Protagonist—Remember that bit about the power of one? There is just one main character that we follow.
4. Active Protagonist—The main character actually does something, doesn't just sit around thinking and agonizing.
5. Closed Ending—The story is resolved; all questions are answered.

The Wizard of Oz is a classic example of archplot. It has (1) a linear timeline (doesn't jump around in time); (2) causality (we get to see the reasons for everything—the flying monkeys don't appear out of nowhere, they are sent by the Wicked Witch); (3) a single protagonist (Dorothy); (4) an active protagonist (Dorothy is actively following the Yellow Brick Road, not just sitting, waiting to be rescued);

and (5) it comes to a closed ending (Dorothy finds her way back to Kansas and lives happily ever after).

It's not a coincidence that *The Wizard of Oz* both conforms to all these elements and enjoys enormous and eternal popularity. The brains of the masses are geared to archplot. Movies that are popular with the masses succeed because they match this structure.

The third corner of McKee's Triangle, *miniplot*, is basically the opposite of archplot. Miniplot minimizes the importance of plot, instead focusing more on character. Take all the archplot traits, reverse them, and you have miniplot: (1) nonlinear timeline (jumps around in time); (2) limited causality (things can happen for no clear reason); (3) multiple protagonists; (4) inactive protagonist (can't even decide if he wants to fight the bad guys, just sits and agonizes) (5) open ending (the bad guys are never destroyed, the murder is never solved, the guy never gets the girl).

Miniplot films typically play in art houses, are cherished by movie critics, and often win Academy Awards. 1996 was known as the year of the art film as movies like *Shine*, *Fargo*, and *Secrets and Lies* competed well at the Oscars. These types of films garner critical praise but tend to play to smaller audiences.

Looking at these traits helps you appreciate the genius of Quentin Tarantino's *Pulp Fiction*, a movie that conformed mostly to miniplot (multiple protagonists, incredibly nonlinear, a fair amount of randomness) yet connected with the mass audience. (In his wonderful book *Into the Woods: How Stories Work and Why We Tell Them*, John Yorke analyzes *Pulp Fiction* in depth and shows that, in the end, the movie's structure is actually quite conventional.) Tarantino included enough elements of archplot (eventually closed the ending, very active protagonists, plenty of external conflict) to reach the masses and enough miniplot characteristics to wow the art critics. The movie fell somewhere between the two corners—artistically respected, yet broadly popular, but still not on the list of the top 50 all-time box-office earners.

For our discussion of movies, think of these three pure story forms as mass entertainment (archplot), art house films (miniplot), and "who cares about the audience" (antiplot).

Now think about how many art house cinemas are left in our society. A few. Not many. Most people prefer archplot. The archplot structure has become programmed deeply in our nature over the ages. Consequently, if you are telling a story of science, as soon as you start violating those basic attributes of archplot, you're starting to lose people.

1. As soon as you're telling a story of science and you start jumping around in time, you're losing people.
2. As soon as you're telling a story of science in which things happen for no clear reason, you're losing people.
3. As soon as you're telling the story of several scientists or projects (multiple protagonists) instead of just one scientist or project, you're losing people.
4. As soon as you're telling a science of story with internal conflict (should we even do this experiment?) rather than external conflict (actually doing the experiment), you're losing people.
5. As soon as you're telling a story of science with no ending, you're losing people. (Ringing any bells here climate change people?)

None of these violations are lethal to the story you're telling, they just come at a price. Commit enough of them, you'll have a small audience.

Bringing It All Together: Archplot, Positive Results and the ABT

Now let's make some interconnections between science and movies using McKee's Triangle. There are three potential outcomes of a scientific study: (1) you reject the null hypothesis, leading to the conclusion of a positive result; (2) you find you are unable to reject the null hypothesis, leading to the conclusion of a null result;

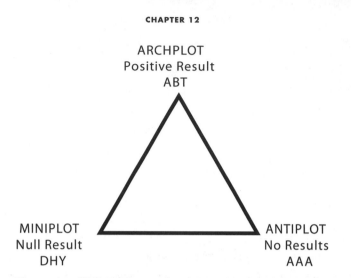

Figure 14. The merging of Hollywood, research and communication. McKee's Triangle broadened to include the outcomes of scientific research and narrative structure.

(3) you fail to gather enough data to draw either conclusion. The positive result is archplot, the null result is miniplot, and not enough data is antiplot.

Think of what this means. The positive result conforms to the exciting, broadly appealing, widely interesting story dynamics of box-office successes. The null result gets the same respect as an art film, but grabs the interest of a much, much smaller crowd.

This is the narrative dilemma and is the reason that, as a scientific researcher, you need an intuitive feel for narrative. You think you're conducting a humble research project and that the scientific community will examine your subsequent report with clinical, robotic, dispassionate scrutiny. But whether you realize it or not, your project has this cloud of narrative energy attached to it. Not all outcomes are created equal.

This is something you must be aware of. If you aren't, you will be subject to all sorts of problems. You may be subconsciously drawn to overreaching with your conclusions—wanting to capture the largest possible audience with archplot dynamics. Or you will be stunned and disappointed when you find out that no journal is willing to publish your null results—welcome to the world of miniplot.

There is one further connection we can make in terms of narrative content. Archplot is the ideal form that the masses yearn for. It is the same as the ABT. Miniplot is the sophisticated, intellectual, complex form that is often more highly respected, but too much for the masses. It is the equivalent of DHY. And antiplot is the "I don't care about the audience" attitude that is matched by the AAA structure of just presenting data with an attitude of "Let the audience figure out for themselves what it means."

So there are the three corners of McKee's Triangle connected to the world of scientific results (the scientific method) as well as the communication of science (narrative structure).

Swimming Upstream against the River of Story

I have told you about McKee's Triangle in an effort to help you see how ancient, mammoth, and inertial the flow of story is. No one is immune to this power. We are all caught up in it. We are all drawn

Figure 15. The truth swims upstream. What happens when the truth is battling story dynamics? Will it get swept downstream?

to archplot. Religions build their entire existence upon it, telling stories that conform to the basic dynamics of archplot—single protagonists, actively and in a linear manner overcoming obstacles and learning lessons until finding their way to a closed ending.

The power of archplot can be like swimming in a powerful river. Which is okay when the river is flowing in the direction of the truth. But what about when it isn't?

Henry Fonda Was a Scientist at Heart

What happens when the truth is forced to swim upstream against the river of story? I'm afraid we all know about the problem of lynch mobs. They are nothing but a river of story headed in the wrong direction, but so powerful that the truth has no chance against it.

A great lynch mob story is *The Ox-Bow Incident*, a novel that became the 1943 Oscar-nominated classic western starring Henry Fonda. It's the story of the few fish who believe in the null story (the bad guy didn't do it) trying to swim upstream against the positive story of the mob, who are certain they have their man (keeping in mind that by "positive" I mean "positive result" in the scientific sense).

Overall, the movie is a null story. It is not an archplot tale of heroism. It is not *High Noon*. It is not a tale of confronting a villain and destroying him with absolute and complete certainty that good has prevailed over evil. Instead it's an open-ended story in which justice doesn't get served and good doesn't prevail over evil.

Not surprisingly, the movie had a hard time getting made. The movie proved to be a "passion project" for Henry Fonda who, at the prime of his career, worked for scale wages and helped raise funding for it. It received an Oscar nomination. Veteran actor Harry Morgan (who was also in *High Noon*), near the end of his career said, "*Ox-Bow* is the best picture I've ever been in." But guess how it performed at the box office. It was a financial dud. Could it have been any more predictable based on the null story at the center of it all? It was pure miniplot.

This is the dynamic that underpins so much of the world of science today. The desire to tell big stories, driven by the river of story, results in false positives, while the lack of interest in telling poor stories (null results) causes the bias against their publication.

This also happens with funding agencies. Too often they aren't interested in funding a test of whether something existing is true or not. They would much rather underwrite a new story that reports a clear pattern. They are driven by that river. I have friends who have been told, verbatim, by funding officers, "We need you to tell a compelling story." To which my friends want to reply, "Oh, we thought you were more interested in the truth."

And it happens with any paper reporting a null result. The editors are matching it up against a positive result paper and, to some extent, estimating the potential audience of each. A term that gets used in the science community is *importance*—as in whether a project constitutes important research. But *importance* in scientific publishing translates mostly as how many people are going to actually care about your findings. Those who are making the decision to accept or reject are caught up in the river of story.

In recent years there have been major efforts to counter this problem. The online journal *PLOS One* (Public Library of Science) was founded specifically for this. The editors' basic philosophy is to accept papers for publication based on "soundness of research rather than significance." In theory this should offset some of the nonpublication bias, but the response of every scientist I talked to about this is "Yes, it is helping, but only somewhat." The draw toward "important" research is relentless, and it becomes the major acceptance criteria for the two most important publications, *Science* and *Nature*.

The divide between archplot and miniplot dynamics can be seen in countless circumstances. Here's an example I heard in a discussion with forest conservation biologists.

Conservation Science: Two Stories

My friends who work with the National Park Service talk about the prevailing philosophy of the government toward conservation in the past versus what it now advocates for the future. In 1916 the US Congress created the National Parks Act with the purpose of conserving "the scenery and the natural and historic objects and the wild life therein and to provide for the enjoyment of the same in such manner and by such means as will leave them unimpaired for the enjoyment of future generations." The key word is *unimpaired*, meaning preserved in the natural state.

There are major elements of archplot embedded in this directive. It implies a closed ending, that the story of nature will end just as it began, with everything the same as when it started, if we can just keep it all unimpaired. This makes for a pretty simple story that is easily conveyed and perpetuated. But unfortunately our world isn't always as simple as we'd like—especially when there is climate change.

In light of this, the National Park Service revised their goals in 2012, recommending the management of parks for "continuous change that is not fully understood, in order to preserve ecological integrity and cultural authenticity, provide visitors with transformative experiences, and form the core of a national conservation land- and seascape."

Whoa. That's a huge dose of miniplot, just starting with "not fully understood." The second element in the archplot list of criteria is "causality," which means that you, the storyteller, understand everything. This is the age-old device of the "omniscient narrator"—meaning that the teller of the story knows everything, especially causality.

But look at what the 2012 statement is saying—basically, "We don't know." That is pure minplot. Furthermore, the Park Service is presenting a story that is continuously changing with no closed ending. What could be more miniplot-ish?

So which of these two "narratives" do you think is easier to explain to the public and even NPS personnel? It's basically a simple archplot story of "we're trying to keep everything as it was" versus the more nuanced miniplot story of "we're modifying our thinking as things change."

For this reason the "preservation" mentality persists in the world of conservation, as does the "balance of nature" concept (the idea that there are invisible forces keeping everything in harmony in the wilds, which religions have traditionally bought into). But neither are accurate concepts for today's world. Such are the frustrations of dealing with poorly programmed brains.

The real world is tough. We can easily find ourselves in the position of that fish—engaged in the upstream struggle against story to establish the truth. I felt it with *Flock of Dodos*. There was a simple compelling story to be told, and a gigantic audience wanting me to tell it. The archplot version of the story would have been that the intelligent design movement was an embodiment of pure evil funded by vicious right-wing tyrants seeking to undermine all of American society. There were many left-leaning folks who not only wanted my film to be this, they even reviewed and promoted it as though that actually was the simple, archplot-ish message of the movie. But it wasn't. As I wrestled to portray the truth of the matter, the movie ended up being closer to the Greeks' concept of the two sides of *agon* rather than the church's portrayal of good versus evil.

Michael Moore (*Fahrenheit 911*), Davis Guggenheim (*An Inconvenient Truth*), Josh Fox (*Gasland*) and other activist filmmakers are much better at the good versus evil thing. And, no surprise, they reach much larger audiences than I could ever dream of. In 2013 I wrote a blogpost titled "Beware the Simple Storyteller: Josh Fox and *Gasland*." It was in response to watching *New York Times* journalist and friend Andy Revkin try to hold Josh Fox accountable on this very issue of storytelling. In a panel discussion with Fox at the Hamptons International Film Festival, Andy began by saying that the film is "a very good polemic," but the filmmaker fails to see the issue of fracking "through the prism" of the complexity of the issue. Which is exactly right. But his comment resulted in the audience—a bunch of left-leaning geriatrics—literally shouting at Andy. (Tough crowd in the Hamptons.)

Anyhow, at this point I should say something bitter about how rich these filmmakers are and how broke I am, but the fact is I enjoy all their films and think very highly of Michael Moore despite his occasional lapses with the truth. My brain is as faulty as the rest of the masses.

The important point to all of this is that there is this powerful river of story that tends to be archplot in structure while too often the truth is more miniplot. This is how Hollywood looks at the world of narrative, but I'm recommending now that the science world consider the same approach to understanding the narrative dynamics of important science stories. Let me show you what I mean with a couple of real world examples.

Why Global Warming Is Bo-ho-horing

Most climate scientists agree with the idea that humans have altered our climate (87 percent according to a 2014 poll by the American Association for the Advancement of Science). This means there is a positive result to communicate to the public. Seems like it should be easy. Yet a 2014 Pew Research Center poll shows that half the American public still doesn't really buy into it, and worse, they're just not that into it. Climate tends to rank as a low priority for voters' concerns in one poll after another. Why is this? I've got plenty to say about it.

In 2010 Andy Revkin posted an item on his *New York Times* blog *Dot Earth*, the headline for which quoted me as saying that the entire subject of global warming is "bo-ho-horing." In 2013 a group of journalists at *Der Spiegel*, preparing to attend the next big global climate meeting (and dreading it) came across that blog post. My use of "bo-ho-horing" seemed to speak to their hearts, so they interviewed me, asking what prompted me to coin such a term. In the United States, ABC News picked up the interview with the same word in their title. A number of my climate scientist friends were not amused.

I saw the bo-ho-horing stigma coming as early as 2002 when I met with a group of professors at the University of Washington to talk about the oceans, but instead they steered the discussion to their concerns about this impending issue of having to teach about global warming. This was four years before the release of *An Inconvenient Truth*, but they already were seeing the signs of how painfully dull the subject is.

They told me their students hated the subject. They found it boring. Why was that? Much of the answer lies in what we have covered in this book. Let's look at a few of the key factors.

The Miniplot Nature of Global Warming

The "story" of global warming is steeped in miniplot traits. Look at the main attributes we've reviewed.

LINEAR TIMELINE—Is there any clear sequence of events we can focus on to feel a clear build over time toward any sort of narrative climax? Not really. The "story" of global warming is all over the map. In 1988, top climate scientist James Hansen testified to Congress during that exceptionally hot summer and said it was time to address the potential problem of global warming, but then the issue pretty much vanished from the news for more than a decade. It popped back up in the nineties with the Kyoto Protocol, but only in an informational way. Hurricane Katrina and the Gore movie brought it back dramatically in 2005/6, but then the hurricanes went away. Super Storm Sandy brought it back again in the United States, but then that concern faded from the news. The story seems to just ebb and flow. There hasn't been a clear narrative build to a climax of the sort you get with archplot. I'm not saying there needs to be, only that this is a narrative characteristic of the issue.

CAUSALITY—For those in the know, the difference between weather and climate is as simple and clear as the difference between short-term and long-term patterns. But it's not automatically obvious to the public. The result is an appearance of randomness in the patterns, which implies a lack of causality. One heavy winter snowfall or cold snap seems to send a signal contradicting the whole idea of warming. When larger-scale predictions do not come true, the pattern feels especially random. This became painfully clear in the years after 2005 and "The Summer of Hurricanes," in which five major hurricanes hit the United States, Hurricane Katrina being the most memorable. The global warming movement seized on that moment to sound the climate change alarm with a large amount of media attention the following spring around *An Inconvenient Truth*. Part of their message was that "global warming will bring lots of huge hurricanes, just like these!" But 2005 was followed by a string of years in which zero major hurricanes hit the United States, undermining the urgency of addressing the global warming problem and giving a sense of randomness to climate conditions that left the public with a feeling of no causality in the archplot sense.

SINGLE PROTAGONIST—The audience appeal of a single protagonist is pretty much why causes need single leaders. You can come up with all sorts of cognitive reasons why one individual is needed for focus, or you can just address the overall dynamic in saying that one leader fulfills one characteristic of archplot, with all its attendant properties for the masses. So for global warming, who would that one leader be? Al Gore sort of was for a while, but he was not a scientist fighting a battle, nor was he really even fighting a battle, and eventually he dropped out of sight on the issue—with particular finality in 2008 when he said he was shifting his focus to energy. He also lost major credibility by selling his Current TV network to Al Jezeera, which, regardless of the truth (it's the old fish in the river of story), is perceived by the general public as the network either of terrorists (as David Letterman pointed out to him) or the oil interests of the Middle East (as Jon Stewart pointed out to him). For global warming, the protagonists, in theory, are all the "eco-heroes" fighting the good fight to save the planet. Which instantly takes us in the opposite direction of a singular narrative thread. If the pain of a single individual is a tragedy, and the pain of a million people is only a statistic, then global warming ends up being a hyperstatistic about the future pain of billions of people. How can an audience possibly connect with that? It's miniplot.

ACTIVE PROTAGONIST—Audiences cheer on the struggling individual, but from the outset global warming has been presented as a story of the masses, who are simply not very active as a unit, being buffeted around by nature.

CLOSED ENDING—Think about the two narratives for the National Park System. The balance-of-nature narrative matches archplot, implying a closed ending—if we can only restore things to their "natural balance." The climate change narrative is miniplot—providing no closed ending. This is what the global warming crowd have had to contend with as they find themselves in a fundamental battle between the forces of mitigation ("We can stop it!") versus the forces of adaptation ("It's too late, time to figure out how to handle it"). As the Keeling Curve continues to reflect the escalating problem of atmospheric chemistry out of balance, the possibility of returning things to "normal" has vanished. There is no simple, closed ending to global warming that can be offered these days. Which makes it pure miniplot—a narrative headed in an unknown direction.

In none of the above commentary am I finding fault in how the communication of the issue has been handled. I'm simply reflecting on how large the challenge has been from the start. But now let me offer a few words on the sad, tragic, ill-informed, naive, and essentially "narratively blind" handling of this massively important issue.

Global Warming: A Miniplot Mess

In a perfect world, in 2002, back when those professors at University of Washington were looking at me with nervous dread in their eyes, things would have taken a different turn. There would have been a national brain trust assembled to look at the issue of global warming in narrative terms, and they would have issued warnings about the miniplot nature of the issue.

This brain trust would have quickly assessed these various elements and come to the realization that we're looking at one huge miniplot crisis. And in response to this they would have issued at least some narrative guidelines on how bombarding the public with information from a multitude of narrative directions would lead to a "miniplot mess."

The global warming issue eventually rose to staggering proportions in the media world, but there was never any sort of sophisticated approach taken to shape the narrative dynamics. In fact, as Matt Nisbet pointed out in his report "Climate Shift: Clear Vision for the Next Decade of Public Debate," there was hardly even any attention given to communication in the first place. He tells about the 2007 report "Design to Win: Philanthropy's Role in the Fight against Global Warming," which was the manifesto issued by the major environmental groups that came together in the wake of the Gore movie. According to Nisbet, in the entire 50-page report, only two sentences address communication, media and public perception.

The closest thing I've ever seen to this entire idea of viewing mass communication of science-related issues in terms of simple narrative dynamics is Kristof's *Outside* article in 2009 that I mentioned in chapter 1. In his discussion of public health education campaigns in Africa, he points at aspects of storytelling, the faulty programming of the brain, and the difficulties that arise from taking too literal of an approach. But his focus is mostly public health issues, not global warming.

Beyond that, I don't see any efforts that manage to combine narrative insight with broad simplicity. You need both. The National Academy of Sciences may (or may not) have pulled together narrative insight with their symposia on the science of science communication, but they've never come close on the simplicity front (in the way that Kristof did). You really need to turn to Hollywood or the advertising world to find that. In fact, Kristof's article leads with "What would happen if aid organizations and other philanthropists embraced the dark arts of marketing spin and psychological persuasion used on Madison Avenue? We'd save millions more lives."

You Can't Rush the River of Story

As a final note on global warming, let me address *An Inconvenient Truth* in terms of the river of story. When you spend enough years around Hollywood, you come to realize it's not the writers, directors and actors who run the show, it's the producers. They are the true voice of Hollywood. The others are just pawns, hired to bring life to their creative activities. Producers pick the scripts, find the money, choose the directors and major stars, and ultimately decide which stories get told.

Even Steven Spielberg can't make any movie he wants. In 2013 he spoke about how his movie *Lincoln* almost ended up on HBO because he couldn't convince the movie studios to make it. The studios are the producers—they control the voice of Hollywood, which in turn runs major elements of mass communication and messaging in our society.

This was exactly the situation for *An Inconvenient Truth*. It was not Al Gore's movie. He was just a pawn in the hands of veteran eco-activist and Hollywood producer Laurie David.

In the summer of 2005 she was in the thick of the crowd of panicked environmentalists who were certain that the five major hurricanes that hit the United States were a sign that "we have entered a new climate world." I heard this repeatedly at Hollywood environmental events that summer. And I eventually heard it from a climate scientist who was in her 2006 HBO documentary feature film *Too Hot Not to Handle*.

This particular scientist told me about how much that film had been dragged down by the scientific oversight committee associated with it. Each draft of the movie was carefully, diligently reviewed for accuracy by the team of scientists. But he told me that, after the five hurricanes of 2005, Laurie David "basically turned to all of us and said, 'I'd like to thank you for your help, but now we have a crisis on our hands, and we can't afford to be bogged down with your critical input, so we'd like to ask you to leave the room as we get to work on a new movie.'"

That new movie, *An Inconvenient Truth*, was filmed that December—three takes with a live audience at Sunset Gower Studios in Hollywood. I had several friends in the audiences who told me about it. It was released the following spring—less than a year from initial conception to release.

The entire film was created in panic mode, which meant that virtually no story development took place. Nobody figured out the one word at the core of it all. Nobody put together a compelling three-act structure that would lead the audience on a journey in search of the answer to a single question.

No, what was produced was an "And, And, And" movie for the ages that simply chronicled the plight of the well-intentioned Gore AND his failed presidential candidacy AND his concern for global warming AND the slideshow he had been giving for years AND then the slideshow itself as he walked through one climate factoid after another.

For starters you can try applying the Dobzhansky Template to the movie's plot. How would the filmmakers have filled it out? "Nothing in global warming makes sense except in the light of . . ." There wasn't a clear theme to the movie, so there is no clear answer for the template.

I know the environmental and science communities flipped over the movie, but my Hollywood filmmaker friends called it bo-ho-horing, as did I. By 2010 it became one of my standing jokes for my university talks to undergraduates. I would ask, "Okay, who'd like to have pizza and beer tonight while watching *An Inconvenient Truth*?" Crickets. Good storytelling results in people wanting to hear the story again and again.

The Global Warming Narrative That Could Have Been Told

There is at least one solidly structured global warming narrative that could have been laid out. The movie could have opened with a first act built around a simple ABT told to an opening photograph of planet Earth with the voice of the heavens saying, "Once upon a time on a small blue planet there was an atmospheric crisis that threatened all of humanity AND by the early 1980s the problem seemed dire, BUT then the nations of the world came together and passed a treaty, THEREFORE today that problem is set to go away."

After telling that ABT, it could have been revealed that the problem was the ozone hole, the treaty was the 1987 Montreal Protocol, and the fact is you hear little about the ozone hole today because it was addressed so effectively.

BUT THEN (now telling an ABT at the larger, movie-long scale with that first ABT serving as the "and" portion) a second atmospheric problem emerged in the late 1980s (global warming) and those same nations that solved the ozone problem have been unable to solve this problem. WHY IS THAT?

And there you go—the "inciting incident"—the central question that sets us off on a journey that Joseph Campbell would have admired. At that point Al Gore could have been brought in to lead us on this quest to understand why we are failing to address the greatest environmental problem ever. And instead of preaching and lecturing, he could have done something that Socrates would have admired—he could have asked questions.

But he didn't. The movie didn't tell a good story. The movie had only tidbits of humor, and those were at the expense of the Republican party (like Gore's joke about how his elementary school science teacher was probably the science advisor of "the current administration"). The only attempts at emotional content were contrived and off topic (the health problems of Gore's sister and son). It was all a big nonnarrative mess that would have bored and frustrated Dobzhansky, as he would have been forced to listen to a sundry list of facts, some of which are interesting or curious, but ultimately fail to paint a meaningful picture.

Different Strategy: Let the Professional Storytellers Do Their Thing

There are two ways to approach the mass communication of information through narrative. Plan A is to make the media yourself and take a chance on finding out the hard way that you're just not that good with narrative (e.g., the Gore movie). Plan B is to simply let the professionals take care of things—you just hitch a ride with them.

The Centers for Disease Control adopted the latter, very wise strategy in the late 1990s to improve the mass communication of their public health information. In 2001 they formed the Hollywood, Health and Society Project (HH&S), a partnership with the USC Annenberg School's Norman Lear Center, negotiated by Lear Center co-founder Marty Kaplan, to tap into the powerful communications resources of Hollywood.

At the core of the CDC's philosophy was not to make their own media but rather to partner with the professional storytellers. They made a deal, providing funding for a number of staff at the Lear Center in return for two things. First, the center would send out the CDC's public health fact sheets to all the writing staffs of the major primetime television shows. They were not to lobby the shows or make themselves annoying by begging them to do an episode of *Grey's Anatomy* on diabetes, Alzheimer's or Ehlers Danlos Syndrome. They were only to make it clear that if the writers needed help in getting accurate information, the CDC was there to help.

Second, if a show were to create an episode based around a health issue using the CDC's information, the Lear Center would then conduct research on how much information audiences managed to retain about the subject after viewing the show.

In recent years I've helped run workshops with the HH&S at the CDC and listened to the impressive list of accomplishments they have achieved through this project. One of the best examples was an eight-episode story arc on the soap opera *The Bold and the Beautiful* in which one of the most popular characters was diagnosed with HIV/AIDS. In the sixth episode they aired a very simple public service announcement during a commercial break in which the actor who played that character spoke into camera urging viewers to call the CDC hotline for HIV/AIDS information. The hotline, which had been receiving a handful of calls a day, was flooded with thousands of calls. Furthermore, the same PSA shown separately, not during the airing of the episode, produced very few calls to the hotline. Such is the mass power of narrative.

The bottom-line lesson of HH&S, and why I am such an avid fan and supporter of their approach, is that they don't try to tell the stories themselves. They respect the enormous amount of skill and time it takes to create good stories that work. You need creative types with the narrative strengths of people like Trey Parker of *South Park*. Everyone needs to get better with narrative, but at the end of the day, you still need to respect the true professionals and take advantage of their skills whenever possible.

Reframing a Null: The Alar Alarm

Given the bias against null results, can a null result ever gain widespread attention? Perhaps you can think of instances when a null result succeeded in capturing a large audience, but when you look closer, you'll probably see a positive pattern at work. Such was the situation in the early 1990s with "The Alar Scare."

It began with a positive result. The chemical alar (daminozide), which was routinely sprayed on apples for a variety of reasons (regulate growth, keep them on the trees), was declared to be carcinogenic in a report released by the Natural Resources Defense Council. The CBS news show *60 Minutes* ran a segment in 1989 about the chemical that helped sound the alarm. Then actress Meryl Streep testified to congress about the need to ban alar for the sake of school kids eating apples. Before things could get too far out of hand, the US manufacturer of alar, Uniroyal Chemical Company, voluntarily withdrew it.

The positive result ("this chemical causes a problem") gained huge attention. But then a major backlash emerged. The apple growers of Washington got angry and unleashed a huge public relations campaign. They filed a lawsuit claiming $100 million in lost revenues, which was accompanied by large numbers of editorials and reports labeling the entire incident as "The Alar Scare," a characterization that persists with many people today who remember the incident.

So how did the backlash crowd manage to generate interest in a null result (their claim that "the chemical doesn't cause cancer")? By piggybacking it on a sort of parallel positive counternarrative that "the environmental crowd is lying to you." Their message had little to do with the chemical—it was all about the intentions they were projecting on the environmental movement.

Notice that the same thing has taken place over the past decade with the attacks on climate science. The opponents to the climate movement are basically trying to argue a null result—either that there is no warming or that humans play no role in the process. But the louder part of their argument is the positive narrative suggesting that environmentalists have a larger agenda at work and are dishonest. All of the opponents' rhetoric is wrapped in archplot in which they are the forces of good going up against the evil environmentalists.

True null results are difficult to communicate. The apple growers' PR people resorted to vilifying the opposition, which is not what I am suggesting. But if you have a null result that deserves attention, consider taking a lesson from the Alar Scare. See if there is a less literal approach that will "reframe" the issue such that you can piggyback your agenda on a positive pattern that will propagate more widely.

(For the record, the environmental crowd wasn't lying about alar. Elliott Negin of the *Columbia Journalism Review* wrote a detailed analysis of the entire incident in the late '90s titled "The Alar Scare Was Real." He conceded, "Like most media myths, this one includes a fact or two," but he went on to pick apart the successful "backlash" campaign that left most of the public with the feeling there never was a health risk with alar, which is incorrect. The carcinogenic risk of alar remains well documented to this day, and it continues to be banned in many countries around the world.)

Yielding to the Narrative Imperative

People need their narratives. This is something that you would do well to absorb, appreciate and respect.

Furthermore, some people seem consumed with the need for certain kinds of narratives, such as the overall story of the world falling to pieces. There is actually a term for this pessimistic outlook—*declensionist narratives*. This outlook is common among environmentalists, to the point where some refuse to even hear good news about the world. They are only receptive to stories of decline—pretty much like Debbie Downer of *Saturday Night Live*.

When you're dealing with people like this, and you need to communicate something contrary to their narrative, you have two options. Either you can tell them they are wrong, or you can roll with their particular "narrative imperative."

Let me share an example of this. I live in a small city on the California coast that, let's just say has a large number of declensionist fans. In 2011, when the nuclear reactor in Fukushima, Japan, exploded, there was an instant wave of fear that the radiation was headed our way.

But four years later a multitude of studies have shown that very little radiation was detectable even in the waters and fish right around Fukushima. Writing for the blog *Deep Sea News*, a group of top marine biologists, many of whom I know and think highly of, concluded, "There are terrible things that happened around the Fukushima power plant in Japan," but went on to say, "Alaska, Hawaii and the West Coast aren't in any danger."

This then is a null result to communicate locally. There are two ways to present it. The first is to ignore "fear needs" of the local audience. This means presenting the one-dimensional story of "The science shows there is no risk of radiation from Japan." To do this is to confront the audience with a message they don't really care to hear as it is not playing to their narrative desires.

But there is an alternate way to present it, which is to put it into the context of other fears. I did this in an article I wrote for our local newspaper titled "It's a Long, Long Way to Fukushima." For the artwork to accompany the article, instead of putting an X over a radiation sign, I made a list of six major threats to California's oceans, such as overfishing and coastal pollution. Buried in the middle of the list was "Japanese nuclear radiation." This one item was crossed out.

I have no polling data on how this played with the readership. I only know that it was an effort to soothe them with their entrenched narrative of the world collapsing before asking them to consider dropping one threat from the list.

People want to feed their narratives. I'm sure Steven Pinker knows this well. He tried to present the null picture with his landmark book *Our Better Angels*, in which he documented the decline of violence in society over the ages, but I have listened to many people just ignore what he had to say as it doesn't fit their "end of humanity" narrative—no point in discussing it.

And lastly, I know all too well how tough null results are to convey. The most important scientific research I did as a marine biologist was a multiyear study on the crown-of-thorns starfish of Australia. The starfish is legendary for its population explosions (or more scientifically termed as "outbreaks" to draw a closer parallel to pest outbreaks). There was a hugely popular story that had emerged which explained why these outbreaks occur. My research suggested that the story, as great as it is, is wrong. Which meant I was promoting a null story—simply saying, "I'm not sure what causes the outbreaks, but it ain't this thing that this guy is pointing to."

My work was published in the best peer-reviewed journals, but it was basically the voice of the party pooper—I was ruining the big exciting story party already under way. I was pulling the rug out from under the party guests. As a result, my work was tolerated for a few years, but as soon as I left science, everyone decided that I had it wrong and went back to celebrating the previous story. Even worse, as I finish this book in 2015, a new study has just been published using methods that were out of date in 1985 but again happily telling the story everyone loves to hear. The bottom line is, even in the scientific literature, the river of story can overwhelm the truth sometimes. Every good scientist knows this.

13

BUT narrative training requires a different mindset . . .

So narrative training is the solution. And in fact, thinking in terms of physical training, I'm coining the term *narrative fitness* as a groovy way to describe the ultimate goal. My friends who teach improv acting talk in similar terms. They think of improv skill as a muscle that you need to condition through multiple, sustained workouts over time. Improv skill does not spring to life in a single session, nor will narrative fitness.

My *Connection* coauthor Brian Palermo, despite all his improv training long ago, continues to do weekly improv "workouts" as a member of the Groundlings, an improv comedy troupe, just to stay in shape. For 14 years he has been one of the six cast members who perform in "The Crazy Uncle Joe Show" every Wednesday night, week after week. He is a brilliant improv actor, but he will be the first to tell you it is only because he maintains his "improv fitness" so consistently.

I'm recommending the same thing for narrative, but here's what's different—I don't see anyone teaching narrative from this perspective. Many books and workshops talk about story. They bring in great experts on storytelling, they run countless one-day or even weekend workshops on storytelling, but none of the thinking is along these lines of fitness training. This is a problem.

If you want to really learn this stuff, it's going to take more commitment than doing a one-day workshop. One day is enough to learn the basic rules and commit them to memory, but very little is going to happen at the visceral level. To grasp narrative at the level where you can just automatically feel it in the material you are reading or working on takes time.

Not only are one-day workshops not enough, there's actually a potential downside to them. I had a nasty teleconference spat with several communications staff at a certain government agency. They had invited my *Connection* coauthor Dorie Barton and me to do one of their large teleconference events with about 50 people who would tune in. As part of the preparation, five of the staff did a half-hour warm-up with us.

Dorie was running through some of what she would be saying about Joseph Campbell and her Logline Maker, but they cut her off. One of the staff said, "We've heard all this before—we've had two of these storytelling workshops, so we know all about the nine elements of the story."

Well, aside from it being the rudest treatment I have ever received from a scientific organization, it was simply wrong. There are no "nine elements of the story." It happens that Dorie had crafted nine elements for her Logline Maker Template, which she had yet to release publicly, but there's a whole range of variations on the number of story elements.

But the real problem was the idea of "we've already heard this."

Increasingly I run into people who have "been there, done that," and "heard" everything in a one-day workshop and somehow think they've got the story thing down. How is it I've spent 25 years working on this stuff and still feel like I'm just getting started, yet some people are certain they have already learned it all in one day? Am I a slow learner?

It's actually more than even Gladwell's 10,000 hours thing. Narrative is a lifelong pursuit that no one ever completely masters. I've

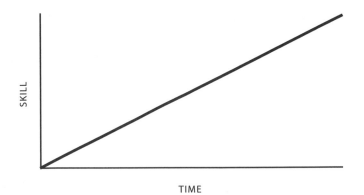

Figure 16. The learning curve for narrative.

put together the simple, almost silly graph in figure 16 to make this point. The graph probably should hit a peak long before the end of a person's life. Few great novelists ended up doing their best work in their last years. Furthermore, today we are dealing with a changing communications environment, which means that the entire dynamic of narrative is probably changing at least a little bit with time.

Who knows. The point is, be skeptical of the advertising hype about "master storytellers." Even Steven Spielberg has had some dud stories. It is endlessly challenging, so all I'm asking is that you not take on the attitude of "Yeah, I've got that story thing down." No, you don't.

It's about "inculcation"—learning by hearing things multiple times in multiple ways. That's the word at the core of effective education. It's at the heart of what a media distributor told me a decade ago about television commercials—people need to see a television commercial four to seven times before they even start to take an interest in what it's about.

Now you might be flashing back to earlier when I talked about the need to "advance the narrative." And you might now be wondering why I am advocating repeating something over and over again. Isn't that failing to advance the narrative?

There is an important difference between inculcation and failing to advance the narrative. If you have a clear central theme that you're seeking to convey that you address from five different angles, you have the potential to achieve effective inculcation. But if you have just one point you're making over and over again the same way, you are failing to advance the narrative. Which is boring. And monotonous. And turns into an "And, And, And" presentation.

To return to the one-day workshops, I worry they aren't helping things much. Yes, they are fun and exciting in the short term, but they are sending the wrong signal over the long term. If you want to learn narrative, you have to make a deep and serious long-term commitment.

Again, it's like a muscle. You can no more take a one-day storytelling workshop and emerge as a person well versed in narrative dynamics than you can complete a one-hour weightlifting workout and expect to walk out buff.

If you're really willing to accept my message of the breadth of scope of this challenge, then you also need to think about the basic dynamics of learning and about how to change an entire system. The ultimate solutions rest not in the cranky old white men who still run the world of science (myself included), but in the next generation whose minds are not yet ossified.

Intuition Starts Early

Let's talk about surfing. For all my years as a marine biologist, surfing was the sport I most dreaded. Over and over again I would visit beautiful ocean settings where the world's best marine biological laboratories exist. But at almost every laboratory a group would eventually go surfing for a day, and I'd feel obligated to join.

The experience would always go badly. It would turn into an afternoon of embarrassment, humiliation and even danger. I would clumsily paddle into waves without a clue of what I was doing, only to be lifted up, thrown over, and pounded down. I remember doing this in Hawaii, North Carolina, Puerto Rico and Australia (though not Antarctica!). It got to the point where I flinched at the mere mention of the sport.

And yet, I love the ocean and was always fascinated by the idea of someday actually being able to stand on a board and ride a wave to shore. So when I moved to Los Angeles for film school, I finally made it a priority to overcome my surfing deficiency. And I did, at age 46. I started surfing every weekend with two buddies—week after week, year after year for more than a decade. I got so serious about it that I moved to the coast, into a house literally within a stone's throw of good surfing waves. Eventually I got good enough to surf a shortboard in the biggest surf with the best surfers without embarrassing myself. But as hard as I've worked at it, I'm not like the guys who started surfing in grade school.

In fact, one of my surf buddies and I were in Nicaragua last year and paddled out into monster 25-foot waves—far too large for me to catch. I felt pretty certain no human could ride them. Until we came around a point and there were a dozen teenagers nonchalantly riding those building-sized waves like they were skateboarding down the hill of their driveway.

And my point is . . . it's all about intuition, gained through experience, which is much more easily and better formed in youth than when you're 46. Youngsters are a blank slate; they don't have to deprogram themselves.

In surfing it's about knowing where to put the board as the wave comes in. Old guys like me, we study the wave and try to figure it out using our brains. But I have watched kids—I'm talking ten-year-olds now—just go for it. They're not even thinking. They're just sensing it already.

This is what the science world needs—young scientists who already, by the end of their undergraduate education, have developed an intuitive feel for the basic rules of narrative structure. They have absorbed the basic principles of this book so deeply that they have a clear understanding of the far-reaching consequences of positive versus null results, as well as the ability to glance at the abstract of a paper and intuitively feel it has too little narrative or too much.

The science world needs scientists who have invested 10,000 hours working on grasping narrative. It is not impossible if you start young. When I first read Gladwell's theorizing about the 10,000 hours, I thought of the greatest storytellers I have ever known, my college buddies at the University of Kansas. Those guys told stories with as much rhythm and flow as an Irishman in a pub. The best had grown up on farms in western Kansas in the 1960s and had clearly been telling stories their whole lives.

I'm certain that, by the time they arrived at college, they were long past the 10,000-hour mark. And it showed as they would silence a noisy bar telling a tall tale of life on the farm (which of course usually involved sexual content). They were storytellers who started young and had truly powerful narrative intuition.

In the science world, introducing narrative early and in a substantial way will produce a whole new breed of scientist, able to communicate far more effectively among themselves, as well as with the public. They will also be less prone to subconsciously reach for false positives or present null results in such a boring way that they help perpetuate publication biases against such results.

But whether you are teaching young people or working toward narrative fitness as a practicing scientist, the same exercise will work. This exercise meshes the core of Hollywood story theory with the doing of science and is my recommendation for solving the problem of narrative deficiency among scientists. I call it Story Circles.

14

THEREFORE I recommend Story Circles

I firmly believe that narrative deficiency is the single biggest problem facing science. You see its cause in the absence of narrative training in science and its effects in the absence of narrative standards within institutions. This absence of narrative standards means you can give science talks that are boring or confusing and no one will say a critical word. I also hear the complaints everywhere I go about how "everything from my field in *Science* or *Nature* today is overstated" or "some of the presentations from our scientists are so bad."

Two things. First, at the individual level, we can use narrative training to build narrative intuition—the long-term solution I am advocating. Second, within institutions, we can strive to create a narrative culture.

Having a narrative culture within an organization or university department or research institution could mean you have reached a critical threshold of people who have undergone narrative training, have developed the basics of narrative intuition, and now the norms have shifted. They know the narrative templates (see appendix 1). They speak with a shared narrative vocabulary (see appendix 2).

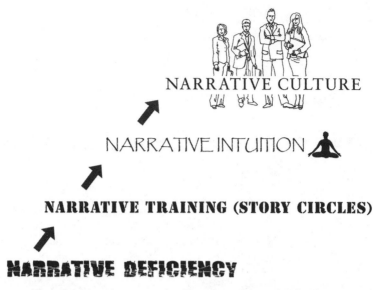

NARRATIVE CULTURE

NARRATIVE INTUITION

NARRATIVE TRAINING (STORY CIRCLES)

NARRATIVE DEFICIENCY

Figure 17. The path to narrative competency. Narrative training using the Narrative Tools in Story Circles leads to narrative intuition. If enough people in your institution follow this path, you will establish a self-perpetuating narrative culture.

And there is now a certain level of expectation of narrative clarity and cohesion.

It's not a lot to learn, there's just a need to learn it well. So to achieve this, it's now prescription time.

Story Circles: A Means of Creating a Narrative Culture

Imagine an institution in which everyone is really good at getting to the point (being concise) and focusing on things that are important (being compelling). Imagine being able to hand back a manuscript and say, "To be honest, it's a bit of an 'and, and, and' piece." And imagine a group sitting down at the conference table to start a discussion by saying, "Here's our problem—we've got a miniplot situation with this current project we're putting together."

It is possible. It can emerge as a norm where people who fail to practice good narrative structure are given polite hints at how to work on it. And it can produce "entrainment"—the process through which others are pulled along in a direction by the movement of the group.

If an institution achieved this situation it could be said to have established a narrative culture. I have yet to encounter such an institution, but I've talked to several administrators who dream of their own institutions being such places. At one major aquarium I led a two-hour discussion among some of the staff on exactly this question: "How can we establish a narrative culture here?"

It begins by having the tools, which then need a means of implementation. But what I am advocating is a fundamental departure from the usual philosophy toward propagating this information.

The standard approach to teaching storytelling consists of shoveling large and exciting amounts of information at workshop participants. This is the sort of experience that sends them home with a feeling of "Wow, my brain is about to explode I've learned so much today!" Yet a week later they will often remember very little or will be confused about much of what they recall.

I had a painful personal experience with this. I made a documentary feature film about a part of World War II that involved my father. Working with an editor, despite all my great and mighty wisdom, we created a combination AAA/DHY film. We did test screenings of 30 people in a small theater. They filled out feedback forms. They raved about the movie—the highest scores I've ever gotten in testing of my films. And in the post-screening discussion, several times people said, "This film is packed with so much and plays on so many levels I'm going to need a few days to really process it."

We thought that was a compliment. It took a year to realize it wasn't. Instead, it was a huge red flag that there was too much going on, the film was too complex, it was too dense with information, and it was too episodic in structure ("and then, and then, and then . . ."). This stuff is really, really tough. I'm still working on that film. Complexity can be exciting in the moment, but deadly in the long term.

The approach I'm advocating consists of teaching only a few fundamental items, then having everyone practice, practice, practice them. When they go home from learning the fundamentals, their brains are not feeling that overloaded, but by revisiting the material, week after week, they are able to recall most of the small amount of information they were given plus develop the deeper intuitive feel for it.

You can see why I have compared this process to physical fitness training. The end result of this narrative fitness process is narrative intuition.

Story Circles for the Science World

How can institutions implement narrative training (the practice, practice, practice part of the process) within scientific programs? Based on my experiences in the film world, I recommend creating small "Story Circles" as a means to engage in the cultivation not just of stories themselves but narrative skills in general, eventually leading to the development of narrative intuition. I've implemented Story Circles at major universities, and the participants are seeing success.

At the core of the Story Circle process is "story development," which is what goes on all day, every day in Hollywood. Countless scripts get written each year. Most need "further development." Many end up in "development hell," where they keep getting written and rewritten, year after year. If you ask a screenwriter what happened to that great script of theirs, they may well tell you it's in "turnaround" at best, or at worst "development hell," meaning it may never see the light of day as it is periodically revisited then reshelved.

Now, of course, I'm not talking about getting so deep into the weeds of this screenwriting stuff that you end up emulating the Hollywood process. But what I am talking about is starting with the realization that story development is how you achieve narrative intuition, and story development works best as a group process—thus Story Circles.

In film school they programmed this need for group dynamics into us from day one. They hit us on the first day with the dictum "Film is a collaborative medium." This means you can try to make an entire film all by yourself, but you're being foolish if you do, and you'll probably end up with something that feels a bit off—a little mutated and not quite right. Basically you'll end up with the product of a "jack of all trades, master of none."

If you're talented, your own "voice" can probably take things about 70 percent of the way, but then you need the input of others. They help you see the blind spots. They reveal the misconceptions (things you thought were funny actually aren't to an audience; things you thought were dramatic turn out to be funny to an audience).They help you achieve a broader voice that will reach the wider demographic.

You get all these benefits by working with other individuals. It's not an easy process, but it is essential. The process also fosters collaboration, a basic skill you will need as a scientist for the rest of your life.

So here is my initial suggestion: create small groups that will meet regularly to practice the basic elements of narrative. I would recommend five people meeting weekly if possible, and attempting only ten weeks to start with. The first few meetings may seem only marginally useful, but it's the repetition, not the individual meetings, that will eventually make the difference.

First the group will probably hit a point of tedium and boredom, feeling a bit of "this again?" But eventually there will come "breakthrough sessions." What makes me think this will happen? The intensive two-year Meisner acting program I took when I started film school, which ended up providing the core of the ideas for *Don't Be Such a Scientist*. Before I started the program, a previous graduate warned me about this exact experience. She said, "You will drive home some nights frustrated, bored and even depressed. But you will also drive home some nights saying, 'I get it,' as you feel exhilarated, inspired and on fire."

Everything she said came true for me. Some nights were torture, but the breakthrough nights definitely happened. There were those specific moments of discovery—almost like the Invention Ambassadors—where I just gazed off in the distance and said, "Wow, I get it." That's how the building of genuine intuition works. Not in an instant, but rather slowly over time, incrementally. And with up and down variation. It's what happens with repetition.

Actors Know about Intuition

If there's one group of people who know how to turn analysis into intuition, it's definitely actors. The essence of bad acting is the delivery of a cerebral, rather than intuitive, performance. As my crazy acting teacher screamed at the class, night after night, "You're too caught up in your head!" Especially me, the former academic.

The repetition approach of the Meisner technique is legendary in its effectiveness. You can hear it in the testimonials of countless actors, from Gregory Peck and Grace Kelly to Michelle Pfeiffer and James Franco.

So I recommend breaking each week's hour session into two parts, somewhat like the *They Say, I Say* divide. The first half, to get warmed up, consists of "they say." Participants use the Narrative Spectrum to analyze abstracts from scientific papers. The abstract is shared as the group looks at its structure and determines where it lies on the Narrative Spectrum from AAA to ABT to DHY. With each diagnosis, intuition for narrative structuring drifts a few nanometers deeper into the visceral realm.

The second half is the "I say" element, where one group member shares his or her current "story" project. The story could be a real story of the individual's journey through a research project, or it could be the story of an entire research program. Rather than giving a full presentation, the individual simply presents the ABT version of the project, study, program or whatever is being worked on.

As the individual goes through the ABT, everyone in the group listens and does his or her best, not to critique or criticize, but to ask the essential questions like "What's at stake?" "Is this the optimum composition for balancing concision with being compelling?" "Is it too general to be powerful?" "Does the narrative move along?" "Is there any emotional content that could be explored?"

The last question leads to some of the elements of the Paragraph Template—"Is there a taking-stock moment that could be developed more deeply?" "Is there a flawed protagonist dynamic that could be further brought out and taken to term?" "Is there a darkest hour?"

Once the ABT is firmly established, then the group moves on to the Word element using the Dobzhansky Template. Is there a single word or phrase at the core of the story? There might not be, but first, you'll never know unless you think about it, and second, you still might not get it unless you think about it aloud while others are there to push you.

An opening, one-day workshop is only a tiny start in the right direction. I can come in and excite everyone for a day, but a week later much of what I taught will evaporate and little will remain. But if the work continues in Story Circles, forcing you each week to dive back into these core principles in a practical way, helping each other, the material will begin to make the move toward intuition. That is the ultimate goal of everything I have to offer—to strive for excellence and even perfection.

And will it get boring? Not if the content continues to be fresh. I ran a workshop with the Society for Marketing Professional Services where I had 30 participants sit in a circle and read their ABTs aloud. I had wondered whether by the time we got to person number 26 everyone would be feeling like, "Enough of those three words." But that never happened.

It never got boring because each ABT was a new and interesting story. The three words, being just the scaffolding for the structure, remained invisible, overshadowed by the content. Once again, there's the power of story.

Can the Shift to Narrative Training Happen Overnight?

So have I lost my senses? Have my 20 years in the Hollywood insane asylum left me devoid of common sense? Do I really think I could make an impact on the problem of narrative deficiency in science?

Maybe. But the one thing I can guarantee is that it will take time. Science is an incredibly conservative profession run by committees and a peer review process that keeps a tight rein on novelty and innovation.

Just how slow is science to change? Philosopher of science Thomas Kuhn described one model of change in science in his 1962 landmark book *The Structure of Scientific Revolutions*. He talked about "paradigms"—existing ways of thinking in science—that solidify and become slow to move or change. However, he also noted the pattern of "paradigm shift," where enough evidence accumulates in one direction to reach a sort of tipping point. When that happens, change can be rapid.

Now let's look at how the science world has changed with regard to narrative. Have the changes come about quickly? When a good idea is introduced, does everyone say, "Hey, good idea, I think I'll change!" Or have the changes been slow and steady? (Basically the same as the classic question for evolution—Does it happen gradually or in short, rapid bursts?)

Time to jump back to the start of this book and look again at figure 1 (on page 7). We are now coming full circle in our journey.

Check out the shape of that plot. What does it tell us about how the change occurred over time? Does it show 40 years of resistance then a "tipping point" where it all took off overnight? Or was there an immediate adoption of IMRAD followed by 40 years of battling for complete acceptance?

No, I'm afraid what you see in that graph is a painful portrait of the conservative nature of science. It is almost a straight line—a slow, steady grind. There's maybe a trace of an inflection point in the middle, but not much. I'm guessing it was just year after year of editors reluctantly giving in. Or more likely, editors convincing their advisory boards to give in. Sadly, what it means for me is that I can probably expect to see complete adoption of the ABT Template for abstracts somewhere around my 110th birthday.

Creating the Perfect Scientist

So what do I envision when I think of the "perfect scientist"? I spend so much time critiquing and criticizing scientists and the science world that some people ask, out of frustration, "So, Mister Expert, how exactly do you want us to be?" Fair enough. I have a clear answer, which I will break into the two dimensions around which we have structured our Connection workshop—the cerebral versus the visceral.

For the visceral side of a perfect scientist, I offer up what I saw with the AAAS-Lemelson Invention Ambassadors. They are role models—very smart, very disciplined, very critical scientists who also are extremely creative, very broad thinking, able to listen effectively, capable of very nonliteral thinking, and most important of all—they get along with other people really well.

I have known thousands of scientists over the years and have worked with hundreds of them in my workshops over the past decade. The six scientists I got to know in that exercise were different. Here's how.

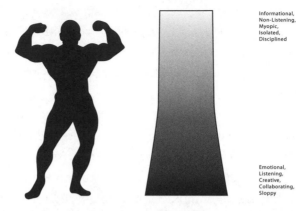

Informational,
Non-Listening,
Myopic,
Isolated,
Disciplined

Emotional,
Listening,
Creative,
Collaborating,
Sloppy

Figure 18. The perfect scientist. The perfect scientist is strong at both ends of the spectrum—in touch with creativity, the ability to listen and a sensitivity for human content, yet still retaining the fierce, incisive and disciplined mind ultimately needed to produce science that will stand up to critical review.

For starters, they embodied everything that Brian Palermo teaches in the improv portion of our workshop. They listened. Unlike most scientists, they listened and did their best to affirm (not negate) input. I saw this in detail when I gave them notes on their talks.

They collaborated. I saw this the first day from the moment they entered the room. You could see how interested each one was in the other members of the group. They asked questions of each other and actually listened to the answers. No one was trying to impress the others or "one up" anyone. They were incredibly humble.

They were also, not surprisingly for inventors, so creative. When I offered up a suggestion on their stories—something like, "Could you tell us something that gives an idea of how you initially got interested in nanotechnology?" instead of negating ("Why would I want to tell about that—nobody is going to find that interesting"), they heard what I said then replied with, "What if I told about this . . . or what if I told about this . . ." That is the "yes, and . . ." dynamic of improv at work, even though none of them had gone through improv training.

At one point I felt as if I were actually in an improv class. One standard exercise for improv training is "next choice," in which the participant says something like, "So I bought a new car . . . ," and the instructor shouts, "Next choice!" and the actor offers up a new choice (usually escalating upward), saying, "So I bought a new truck . . ." "Next choice!" "So I bought a new tank . . ." A good improv actor can spew out next choices instantly, without hesitation. A bad one pauses, thinks, and gets locked up.

As I asked for new content, instead of locking up or negating, they flipped into idea mode and were able to offer up one suggestion after another. In every facet they embodied the traits of improv.

It's not a coincidence. Improv fosters creativity; inventors are highly creative. Creativity requires that you "get out of your head"— that you shut down the negating machinery that comes from the cerebral end of the spectrum. These are people who have lived their lives driven by their creativity—looking at problems and coming up with creative, nonliteral solutions—making intuitive leaps and exploring countless ideas rather than shutting things down and being myopic.

My working with them changed my perception and optimism when I think about scientists in general. I had thought it might be possible to be a good scientist *and* be that creative, but I'd had my doubts. No longer.

The other asset of a perfect scientist is the core message of this book—narrative intuition. I don't have the time and resources to poll all the recent Nobel Laureates, but I'm guessing you would see strong narrative intuition among them such as I have noted for Randy Schekman and James Watson. It's not a coincidence.

The perfect scientists would be capable of more than just communicating with great narrative structure. They would also have such a strong understanding of the narrative side of the scientific method that they would have an impact on the existing problems of false positives and underpublished null studies.

All these traits are possible in a single scientist and are present in many of the greatest ones today. They just take extra effort to cultivate, which is why they need to be taught and encouraged from the very start of a science career. Excellence is possible, but only with proper training, and the right overall perspective. Which brings us back to the big picture.

This View of Life

Stephen Jay Gould was the greatest scientist I ever met. For me, he was the embodiment of the perfect scientist—at least in the early years of his career, the late 1970s, which was when I was fortunate enough to spend time around him. He was steeped in both the cerebral (a member of the National Academy of Sciences by age 48) and the visceral (full of humor, deeply impassioned). But his professional life ended up being a sad and important story that underscores all I have presented in this book as he eventually impaled himself on his own sword.

Gould spent his entire career warning of the human failings of scientists. For 25 years he wrote a monthly column in *Natural History* titled This View of Life in which he told the tales of Piltdown Man, "The Case of the Midwife Toad," and even argued for respecting Lamarck, the guy before Darwin who is usually laughed at for getting evolution wrong. In many ways it was his life's work— saying you can't understand the scientist without understanding the human weaknesses of the scientist, for which the desire to tell big stories is perhaps the greatest.

In 1978 (the year I first met him) at the end of the abstract of a paper he published in *Science*, he warned of exactly this, saying, "Unconscious or dimly perceived finagling is probably endemic in science, since scientists are human beings rooted in cultural contexts, not automatons directed toward external truth." But here's the sad part.

A few years after he passed away, others determined Gould himself had fallen victim to this failing in what was perhaps his most important popular work, *The Mismeasure of Man*. In this book, Gould accuses 1800s physician Samuel George Morton of having racist biases in the way he measured the cranial capacity of skulls. I actually remember Gould excitedly telling us in his Tuesday lunch group about his discoveries of Morton having "fudged" his data. You could see in his eyes that he had discovered "a good story."

In 2011, two anthropologists reexamined Gould's analysis and found that it was actually Gould who was guilty of "confirmation bias"—the tendency to make the data support the story you want to tell. In their reporting of this they concluded, "Ironically, Gould's own analysis of Morton is likely the stronger example of a bias influencing results." It isn't clear whether Gould did this knowingly or not, but it is clear he biased his work in the direction of a false positive.

This is the only report of Gould ever committing such an error, but it shows that if even the greatest of scientists can fall victim to the narrative programming of the brain, then every scientist needs to be aware of this human weakness and avoid the ways in which it can undermine the scientific process. Scientists should accomplish this not by banishing story from science but rather by looking it directly in the eyes in order to fully understand it.

Therefore . . . (a fitting word for ending this book) . . . in the spirit of the title of Gould's monthly column, "This View of Life," I hope I have brought you around to at least a slightly different view of life—now looking at things a bit more from the perspective of narrative. It would be great if you began looking at and listening to "stories" of all sorts and seeing if you can feel where they fall along the narrative spectrum. You hear ABTs every day on the news. One NPR story I heard recently began roughly, "Reflective glass on office buildings is environmentally efficient AND has become very popular, BUT the reflected sunlight can cause problems for neighboring buildings, THEREFORE . . ."

I also hope that you will look at scientific studies and at least ask whether what you're seeing is an archplot/positive result/ABT structure versus a miniplot/null result/AAA snoozefest (which, frustratingly, may sometimes better embody the truth). Once you make that determination, a whole new world of dynamics opens up.

If you get yourself to the point where you can spot this distinction instantly, then you're probably hitting the level of narrative intuition that will keep you from making the errors I've warned of for both research and communication. If you can make this one change in your perspective, it could conceivably change your entire view of life. And one day, as you look back at planet Earth and figure out how you are going to explain to your fellow Earthlings what you are headed off to do in a galaxy far, far away, you can call back and say, "Houston, we have a narrative!"

Appendix 1
The Narrative Tools

Word

THE DOBZHANSKY TEMPLATE

Nothing in _____ makes sense except in the light of _____.

Sentence

THE AND, BUT, THEREFORE TEMPLATE (ABT)

_____ and _____, but _____, therefore _____.

Paragraph

THE LOGLINE MAKER TEMPLATE
(AS DEVELOPED BY DORIE BARTON)

In _an ordinary world,_ _____

a flawed character _____

has _a catalytic event_ _____

which upends his/her world, but after _taking stock,_ _____

the character decides to _take action,_ _____

233

but when *the stakes get raised* _____

the character must *learn the lesson* _____

in order to *overcome the opposition* _____

and *achieve the goal*. _____

The ABT Words

These are words that can be interchanged with the ABT words.

Agreement	*Contradiction*	*Consequence*
AND	BUT	THEREFORE
also	despite	so
equally	however	thus
identically	yet	consequently
uniquely	conversely	hence
like	rather	thereupon
moreover	whereas	accordingly
as well as	although	as a result
furthermore	otherwise	henceforth
likewise	instead	for this reason
similarly	albeit	in that case

Appendix 2

Narrative Vocabulary

Narrative Spectrum

AAA—"And, And, And," meaning nonnarrative in structure

ABT—"And, But, Therefore," the optimal narrative form

DHY—"Despite, However, Yet," meaning overly narrative, too many directions

Dobzhansky Template—a tool for finding the theme/core/message

pile of sundry facts—what you get with the AAA structure

failure to create a meaningful picture—what happens without a theme

frame—the context into which a series of events are placed

narrative—the events that occur in the search for the solution to a problem

Narrative Spectrum—the range of narrative structures from none to optimal to excessive

null results—scientific findings that fail to show a clear pattern

positive results—scientific findings that show a clear pattern

theme/core/message—the element that gives overall meaning to a narrative

Story Cycle—the 12-part breakdown of the Hero's Journey as first described by Joseph Campbell

Logline Maker—another template for a story, derived from Blake Snyder's "Save the Cat" teachings

McKee's Triangle

McKee's Triangle—a device for conceptualizing the narrative structure of stories, from Robert McKee's book *Story*

archplot—the ancient, broadest form of story structure, which reaches the largest audience (e.g., blockbuster movies)

miniplot—the opposite of archplot, which minimizes the importance of plot versus character (e.g., art house movies)

antiplot—the dismissal of plot entirely (e.g., "artsy" films)

Appendix 3

Twitter "Stories"

Is the length of a tweet ideal for narrative structure?

The ABT provides one means of addressing this question. For our Connection Storymaker Workshop we have the participants bring a story to work on. In preparation, we have them come up with the ABT for their story, which they email to us in advance. When we counted the number of characters in each ABT, then calculated the average lengths for different groups, the values were more than double the 140-character length of a tweet, as you can see in figure 19.

So if the most basic unit of narrative structure averages twice the length of the longest possible tweet, how exactly did Twitter arrive at the number 140? Was it decided by an international commission of narrative experts? Did the greatest minds of communications come together to determine the optimum length for conveying short narratives?

Nope. Twitter was derived from SMS texting, which is built around a length of 160 characters. As Mark Milian reported in the *Los Angeles Times* in 2009, it was German computer pioneer Friedhelm Hillebrand who in 1985 decided this was an adequate

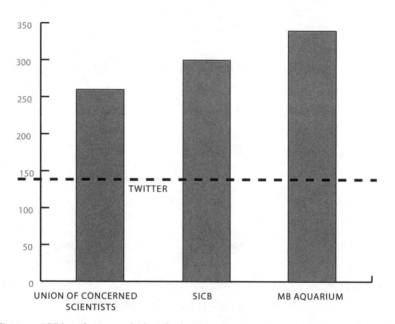

AVERAGE NUMBER OF CHARACTERS IN ABT NARRATIVE

Figure 19. ABT lengths versus the length of a tweet. Each group was at least 20 participants. SICB = Society for Integrative and Comparative Biology; MB = Monterey Bay Aquarium.

length for basic statements. He based this on two main sources— the length of text on postcards (e.g., "Having fun in Stuttgart, wish you were here!") and the length of messages sent through telex (e.g., "Wrecked car, send money now!"). He became the chairman of the nonvoice services committee within the Global System for Global Communications, where he saw that 160 characters was adopted as the industry standard for SMS.

But Stephen Colbert, host of *The Colbert Report*, experienced firsthand the consequences of such a nonnarrative medium. In March 2014 he cracked a joke about the controversy over the Washington Redskins team name that ended with a derogatory name for a fake Asian foundation. The offensive punchline was then tweeted without the setup (i.e., without the narrative context of the complete joke). A firestorm of controversy erupted, to which Colbert responded on his next show, "Who would have thought a means of communication limited to 140 characters would ever create misunderstandings?"

He managed to put the fire out that evening by having on his show Biz Stone, one of the cofounders of Twitter. In their discussion Stone admitted that the maximum length of a tweet is often not adequate to convey a complete thought. In response to this, he and others at Twitter are considering a new version of the app that would allow for longer tweets. I suggest something around 300 characters, the length of an average ABT.

Acknowledgments

I'm a big believer in speaking the truth, especially when it comes to science communication. Toward that end, let me tell you about someone who gave me a painfully honest dose of it.

In the spring of 2010 I met a crazed greenhouse manager named Terry Ettinger during my visit to Syracuse University. As I entered his office he held up a copy of *Don't Be Such a Scientist* with the look of a madman. He began ranting, "You don't understand, I never read books, my friend told me to read this book of yours, I couldn't put it down, and do you know why it was so good?"

I had no answer. He continued, "It's because you told us we need to tell stories full of humor and emotion, and the way you told us was . . . you told stories full of humor and emotion." It was a treat to be so appreciated.

Three hours later, when I finished my talk in the auditorium to the faculty and students, there he was again, in the front row with his hand raised high. I happily called on him first. He stood up, turned his back to me, and gave the exact same wildly enthusiastic endorsement of my book. I looked down from the stage, glowing with pride, soaking it in . . . until he turned back to me and his expression shifted from joy to disgust.

"But now," he said, "I have to just be blunt and tell you that what you presented here to us today was . . . it was . . . boring. You didn't tell us any stories, there wasn't any humor, there wasn't any emotion . . . there was only boring."

Well, the truth hurts. A woman on the other side of the auditorium tried to jump to my defense, but I wanted to say, "Sit down, ma'am, he's right." He really was. I had just completed a total "And, And, And" presentation—lock, stock and barrel—the very thing I'm warning about. It was just a regurgitation of the contents of each chapter of my book, without the humor, devoid of emotion, no stories and, worse, no overarching narrative. Houston, I had a problem.

So I begin by thanking Terry Ettinger for his courage to speak up and call me out. There needs to be more of that in the science world (though a tiny bit more tact might be nice). Finding the narrative in anything is challenging. That's the overall message of this book. Therefore . . . I owe some major debts of gratitude to the key people who have helped me find my overall narrative.

At the top of the list are Bruce Lewenstein, Jennine Lanouette and Jerry Graff. Bruce steered me to the IMRAD acronym, which I knew nothing about despite a twenty-year science career. Jennine steered me to the history of three-act structure, which I knew nothing about despite having earned an MFA in cinema. And Jerry steered me to the DNA of argumentation, which I knew nothing about despite having spent an entire life arguing with pretty much everyone.

"Good Sport" awards go to Mike Orbach and Gary Griggs for bravely and boldly going where few other scientists would dare to venture in working with me on the sea level rise panel for the Coastal and Estuarine Research Federation meeting. In case you're wondering, I did share with them all the details of the version of the story I presented in this book, and they were cool with it. It all really happened that way—including the wonderful and classic, "aren't we special" email. Also, an extra thanks to Megan Baliff for organizing the event.

This book is to some extent a "crowd sourced" piece of work in that so many people attending my talks end up contributing knowledge. One prime example is Tim Olliff, who heard me speak at University of Montana and told me about the National Park Service mission statement I reviewed in chapter 12. Many, many people have made such contributions, including my team of readers, Bec Gill, Jayde Lovell, Jackie Yeary and Steph Yin.

Next come "the usual suspects" to thank, almost all of whom are identified in the acknowledgments sections of my previous two books—basically the same people who helped out back then are still stuck listening to me and providing support. A few others to add to this group are my storytelling buddy Park Howell; my Connection co-creators, Dorie Barton and Brian Palermo; my longtime Hollywood supporters Christina and Fawn; and the person who watched me chop up a million mussels as an undergrad and is still providing me with invaluable guidance in the science world, Dianna Padilla. Also thanks to the journalists I respect most, Philip Martin, Andy Revkin and David H. Freedman.

A couple of important new characters in recent years are my "parking lot soul sistah," the great Shirley Malcom of AAAS, who knows how to see past my rough edges; her co-conspirator Mike Strauss; my Story Circles collaborator Jayde Lovell, who helped shape a number of the thoughts, messages and opinions expressed in this book; and my good friend and constant confidante Samantha, who breaks all molds.

As with my previous book, Vanessa "No Heart" Maynard did a great job with the graphic artwork. Also, a special thanks to Travis Wright for the Cyclops painting.

A major debt of gratitude goes to my editor, Christie Henry, who wrote me such a great rejection letter for my first book that I was eager to try again with University of Chicago Press. She headed up the amazing team behind this book, including Levi Stahl, promotions director and the person who thought up the title (Could I be any luckier? The promotions director has a creative stake in the book!); Isaac Tobin, who designed the wonderful cover artwork; and Jenni Fry, the manuscript editor who actually made this thing readable (and I'm serious).

Beyond that, the ultimate and final thanks goes, of course, to my amazing mother, Muffy Moose—now in her early 90s and still saying the same thing to me she has said all my life: "Shake 'em up!"

Notes

Notes to Part I: Introduction

Page 5   P. B. Medawar, "Is the Scientific Paper a Fraud?" *Listener* 70 (Sept. 12, 1963), 377–78; reprinted in P. B. Medawar, *The Threat and the Glory: Reflections on Science and Scientists*, ed. David Pyke (New York: HarperCollins, 1990).

Page 7   L. B. Sollaci and M. G. Pereira, "The Introduction, Methods, Results, and Discussion (IMRAD) Structure: A Fifty-Year Survey," *Journal of the Medical Library Association* 92, no. 3 (2004): 364–67.

J. Schimel, *Writing Science: How to Write Papers That Get Cited and Proposals That Get Funded* (New York: Oxford University Press, 2011).

Page 9   P. Sumner, S. Vivian-Griffiths, J. Boivin, A. Williams, C. A. Venetis, A. Davies, J. Ogden, L. Whelan, B. Hughes, B. Dalton, F. Boy and C. D. Chambers, "The Association between Exaggeration in Health Related Science News and Academic Press Releases: Retrospective Observational Study," *BMJ* 2014, 349:g7015, http://www.bmj.com/content/349/bmj.g7015.

Page 10   J. P. A. Ioannidis, "Contradicted and Initially Stronger Effects in Highly Cited Clinical Research," *Journal of the American Medical Association* 294, no. 2 (2005): 218–28.

I. Sample, "Nobel Winner Declares Boycott of Top Science Journals," *Guardian*, Dec. 9, 2013, http://www.theguardian.com/science/2013/dec/09/nobel-winner-boycott-science-journals.

Page 11   A. Franco, N. Malhotra and G. Simonovits, "Publication Bias in the Social Sciences: Unlocking the File Drawer," *Science* 345 (Sept. 9, 2014): 1502–5.

R. Olson, *Don't Be Such a Scientist: Talking Substance in an Age of Style* (Washington, DC: Island Press, 2009).

Page 13 R. Olson, D. Barton and B. Palermo, *Connection: Hollywood Storytelling Meets Critical Thinking* (Los Angeles: Prairie Starfish Productions, 2013).

Page 14 M. Harris, "Inventing Facebook," *New York Magazine*, Sept. 20, 2010, http://nymag.com/movies/features/68319/.

B. D. Johnson, "Ben Affleck Rewrites History," *Maclean's*, Sept. 19, 2012, http://www.macleans.ca/culture/movies/ben-affleck-rewrites -history/.

Page 15 *6 Days to Air: The Making of South Park*, directed by A. Bradford, produced by A. Bradford and J. Ollman (New York: Comedy Central Productions, 2011).

Page 16 R. Olson, opening presentation, TEDMED Great Challenges Day, April 19, 2013, https://www.youtube.com/watch?v=ERB7ITvabA4.

Page 17 G. Graff and K. Birkenstein, *They Say, I Say: The Moves That Matter in Academic Writing* (New York: Norton, 2009).

Notes to Part II: Thesis

Page 30 J. Yorke, *Into the Woods: A Five-Act Journey into Story* (New York: Overlook Press, 2014).

Page 33 J. Gottschall, *The Storytelling Animal: How Stories Make Us Human* (New York: Mariner Books, 2013).

Aristotle, *Aristotle's Poetics* (New York: Hill and Wang, 1961).

Page 35 J. Campbell, *The Hero with a Thousand Faces* (New York: Pantheon, 1949).

Page 36 C. Bazerman, *Shaping Written Knowledge* (Madison: University of Wisconsin Press, 1988).

Page 38 N. Kristof, "Nicholas Kristof's Advice for Saving the World," *Outside*, Nov. 30, 2009, http://www.outsideonline.com/1909636/nicholas -kristofs-advice-saving-world.

Page 42 U. Hasson, "Neurocinematics: The Neuroscience of Film," *Projections* 2, no. 1 (2008): 1–26.

A. Gopnik, "Mindless: The New Neuro-Skeptics," *New Yorker*, Sept. 9, 2013, http://www.newyorker.com/magazine/2013/09/09/mindless.

Page 45 S. Terkel, *The Good War: An Oral History of World War II* (New York: New Press, 1984).

Page 48 J. Watson, *Avoid Boring People: Lessons from a Life in Science* (New York: Vintage, 2010).

Page 49 J. Watson, *The Double Helix: A Personal Account of the Discovery of the Structure of DNA* (New York: Atheneum, 1968).

P. Kareiva, "If Our Messages Are to Be Heard" *Science* 327, no. 5961:
34–35.

Page 50   *Flock of Dodos: The Evolution-Intelligent Design Circus*, written and produced by Randy Olson (Los Angeles: Prairie Starfish Productions,
2006).

Page 57   J. Graff, *Clueless in Academe: How Schooling Obscures the Life of the Mind*
(New Haven, CT: Yale University Press, 2004).

Page 58   B. Winterhalter, "The Morbid Fascination with the Death of the Humanities," *Atlantic*, June 6, 2014, http://www.theatlantic.com/education
/archive/2014/06/the-morbid-fascination-with-the-death-of-the
-humanities/372216/.

E. D. Hirsch Jr., *Cultural Literacy: What Every American Needs to Know*
(Boston: Houghton Mifflin, 1987).

M. Slouka, "Dehumanized: When Math and Science Rule the School,"
Harper's, January 8, 2015, http://harpers.org/archive/2009/09
/dehumanized/.

A. Sokal, "Transgressing the Boundaries: Toward a Transformative
Hermeneutics of Quantum Gravity," *Social Text* 46/47:217–52.

A. Sokal, "A Physicist Experiments with Cultural Studies," *Lingua
Franca* May/June 1996: 62–64.

Page 59   C. P. Snow, *The Two Cultures and the Scientific Revolution* (New York:
Cambridge University Press, 1959).

J. A. Labinger and H. Collins, eds., *The One Culture?* (Chicago: University
of Chicago Press, 2001).

E. O. Wilson, *Consilience: The Unity of Knowledge* (New York: Vintage
Books, 1998).

Page 65   C. Vogler, *The Writer's Journey: Mythic Structure for Writers* (Los Angeles:
Michael Wiese Productions, 2007).

B. Snyder, *Save the Cat! The Last Book on Screenwriting You'll Ever Need*
(Los Angeles: Michael Wiese, 2005).

Notes to Part III: Antithesis

Page 74   M. Gladwell, *Blink: The Power of Thinking without Thinking* (Boston:
Little, Brown, 2005).

M. Gladwell, *Outliers: The Story of Success* (Boston: Little, Brown, 2008).

M. Gladwell, "Complexity and the Ten-Thousand-Hour Rule," *New
Yorker*, August 21, 2013.

Page 75   A. Alda, *Things I Overheard While Talking to Myself* (New York: Random
House, 2008).

Page 77 T. Parker, "Funnybot," *South Park*, season 15, episode 2 (May 4, 2011).

Page 81 T. Dobzhansky, *Genetics and the Origin of Species* (New York: Columbia University Press, 1937).

Page 82 T. Dobzhansky, "Biology, Molecular and Organismic," *American Zoologist* 4, no. 4 (1964): 443–52.

Page 91 B. K. Forscher, "Chaos in the Brickyard," *Science* 142 (1963): 339.

Page 92 D. Weinberger, "To Know but Not Understand: David Weinberger on Science and Big Data," *Atlantic*, Jan. 3, 2012, http://www.theatlantic.com/technology/archive/2012/01/to-know-but-not-understand-david-weinberger-on-science-and-big-data/250820/.

Page 95 N. Parsons, "The 7 Key Components of a Perfect Elevator Pitch," *B Plans*, http://articles.bplans.com/the-7-key-components-of-a-perfect-elevator-pitch/.

Page 96 D. Pink, *To Sell Is Human: The Surprising Truth about Moving Others* (New York: Riverhead, 2013).

C. O'Leary, *Elevator Pitch Essentials: How to Get Your Point Across in Two Minutes or Less* (St. Louis: Limb Press, 2008).

Page 101 B. Minto, *The Minto Pyramid Principle: Logic in Writing, Thinking, and Problem Solving* (London: Minto International, 1996).

Page 102 A. B. Lord, *The Singer of Tales* (Cambridge, MA: Harvard University Press, 1960).

F. Daniel, informal talk (edited transcript), Columbia University, School of Arts, Film Division, May 5, 1986, http://www.cilect.org/gallery/news/32/2004-11.pdf.

Page 105 K. K. Campbell, *The Rhetorical Act: Thinking, Speaking, and Writing Critically* (Boston: Cengage Learning, 2008).

H. H. Bauer, *Scientific Literacy and the Myth of the Scientific Method* (Champaign: University of Illinois Press, 1992).

Page 109 J. Watson and F. Crick, "A Structure for Deoxyribose Nucleic Acid," *Nature* 171 (April 25, 1953): 737–38.

Page 111 R. McKee, *Story: Style, Structure, Substance, and the Principles of Screenwriting* (New York: ReganBooks, 1997).

Page 115 A. Greene, *Writing Science in Plain English* (Chicago: University of Chicago Press, 2013).

Page 116 S. Springer, P. Malkus, B. Borchert, U. Wellbrock, R. Duden and R. Schekman, "Regulated Oligomerization Induces Uptake of a Membrane Protein into COPII Vesicles Independent of Its Cytosolic Tail," *Traffic* 15, no. 5 (2014): 531–45.

Page 128 J. Sachs, *Winning the Story Wars: Why Those Who Tell (and Live) the Best Stories Will Rule the Future* (Boston: Harvard Business School Press, 2012).

M. Winkler, "What Makes a Hero?" (video), TED Ed-Original, http://ed.ted.com/lessons/what-makes-a-hero-matthew-winkler.

P. Suderman, "Save the Movie! The 2005 Screenwriting Book That's Taken Over Hollywood—And Made Every Movie Feel the Same," *Slate*, July 19, 2013.

Page 140 C. Keane, *How to Write a Selling Screenplay* (New York: Three Rivers, 1998).

Page 142 K. Weinersmith and Z. Faulkes, "Parasitic Manipulation of Host Phenotype, or How to Make a Zombie," *Integrative and Comparative Biology* 54, no. 2: 93–217.

Page 154 T. Friedman, *Hot, Flat, and Crowded* (New York: Farrar, Straus and Giroux, 2008).

J. Diamond, *Guns, Germs, and Steel: The Fates of Human Societies* (New York: Norton, 1997).

Notes to Part IV: Synthesis

Page 179 K. A. Quesenberry, "Johns Hopkins Finds with Super Bowl Commercials, Storytelling Beats Sex," press release, Johns Hopkins University, Jan. 31, 2014, http://releases.jhu.edu/2014/01/31/johns-hopkins-finds-with-super-bowl-commercials-storytelling-beats-sex/.

Page 180 Y. Katz, "Against Storytelling of Scientific Results," *Nature Methods* 10 (2013): 1045.

Page 182 M. Dahlstrom, "Using narratives and storytelling to communicate science with nonexpert audiences," *Proceedings of the National Academy of Sciences* 11 (2014): 13614–20.

Page 196 A. S. Leopold, S. A. Cain, C. M. Cottam, I. N. Gabrielson and T. L. Kimball, "Wildlife Management in the National Parks: The Leopold Report," *Crater Lake Institute*, Mar. 4, 1963.

Page 198 R. Olson, "Josh Fox and Fracking: Beware the Simple Storyteller," *Benshi Blog*, July 18, 2013.

Page 199 A. Revkin, "Global Warming and the Tyranny of Boredom," *Dot Earth* (blog), *New York Times*, Oct. 27, 2010, http://dotearth.blogs.nytimes.com/2010/10/27/global-warming-and-the-tyranny-of-boredom/?_r=0.

A. Bojanowski, "Filmmaker Randy Olson: Climate Change Is 'Bo-ho-horing,'" *ABC News*, Dec. 26, 2013.

Page 202 M. C. Nisbet, *Climate Shift: Clear Vision for the Next Decade of Public Debate* (Washington, DC: American University School of Communication, 2011), http://www.climateaccess.org/sites/default/files/Nisbet_ClimateShift.pdf.

Page 207 M. G. Kennedy, A. O'Leary, V. Beck, K. Pollard and P. Simpson, "Increases in Calls to the CDC National STD and AIDS Hotline Following AIDS-Related Episodes in a Soap Opera," *Journal of Communication*, June 2004, https://hollywoodhealthandsociety.org /sites/default/files/for-publichealth-professionals/researchand -evaluation/BBHotline.pdf.

Page 209 E. Negin, "The Alar 'Scare' Was for Real; and So Is That 'Veggie Hate-Crime' Movement," *Columbia Journalism Review*, Sept./Oct. 1996, http://www.pbs.org/tradesecrets/docs/alarscarenegin.html.

Page 211 S. Pinker, *Our Better Angels: Why Violence Has Declined* (New York: Viking Adult, 2011).

Page 226 T. Kuhn, *The Structure of Scientific Revolutions* (Chicago: University of Chicago Press, 1962).

Page 231 S. J. Gould, *The Mismeasure of Man* (New York: Norton, 1981).
N. Wade, "Scientists Measure the Accuracy of a Racist Claim," *New York Times*, Dec. 13, 2011.

Page 237 M. Milian, post, *The Business and Culture of Our Digital Lives, from the L.A. Times* (blog), *Los Angeles Times*, May 3, 2009, http:// latimesblogs.latimes.com/technology/2009/05/invented-text -messaging.html.

Index

The letter *f* following a page number denotes a figure.